READY THE CANNONS!

BUILD **W**IFFLE BALL LAUNCHERS, BEVERAGE BOTTLE BAZOOKAS, HYDRO SWIVEL GUNS, AND OTHER ARTISANAL ARTILLERY

WILLIAM GURSTELLE

CHICAGO REVIEW PRESS

Visit www.ReadytheCannons.com for updates, corrections, teacher's notes, and important safety information.

Copyright © 2017 by William Gurstelle
First edition
Published by Chicago Review Press Incorporated
814 North Franklin Street
Chicago, Illinois 60610
ISBN 978-1-61373-445-2

The author and the publisher of this book disclaim all liability incurred in connection with the use of the information contained in this book.

Library of Congress Cataloging-in-Publication Data
Names: Gurstelle, William, author.
Title: Ready the cannons! : build wiffle ball launchers, beverage bottle
    bazookas, hydro swivel guns, and other artisanal artillery / William
    Gurstelle.
Description: First edition. | Chicago, Illinois : Chicago Review Press
    Incorporated, [2016] | Includes index.
Identifiers: LCCN 2016019822 (print) | LCCN 2016020783 (ebook) | ISBN
    9781613734452 (trade paper) | ISBN 9781613734469 (adobe pdf) | ISBN
    9781613734483 (epub) | ISBN 9781613734476 (kindle)
Subjects: LCSH: Toy making. | Handicraft. | Toy guns. | War toys. | Miniature
    weapons.
Classification: LCC TT174 .G869 2016 (print) | LCC TT174 (ebook) | DDC
    745.592—dc23
LC record available at https://lccn.loc.gov/2016019822

Cover design: Andrew Brozyna, AJB Design, Inc.
Cover images: Cannon illustrations, Shutterstock/MaKars; Wiffle ball and marshmallow,
    iStock.
Interior design: Jonathan Hahn
Interior illustrations: Damien Scogin

Printed in the United States of America
5 4 3 2 1

Cannon to right of them,

Cannon to left of them,

Cannon in front of them

Volley'd and thunder'd;

Storm'd at with shot and shell,

Boldly they rode and well.

    —"The Charge of the Light Brigade"
      Alfred, Lord Tennyson

CONTENTS

INTRODUCTION

I've always found things that shoot to be interesting. I'm not saying it's a good thing or a bad thing; it's just the way I'm wired. And since you're reading this book, I'm assuming that you are too.

In *Ready the Cannons!*, I want to explore with you the world of things that shoot. We will do that in three ways—first by delving into the science that makes projectiles project, then by exploring the history of big guns and the people behind them. And last but by no means least, we will go hands on, put hammer to nail, drill bit to metal, sandpaper to wood, and build some pretty incredible guns, cannons, and other, more exotic (shall we say weirder?) shooters with our own hands.

I've spent a considerable amount of time choosing the mix of projects. Some of the projects are big guns indeed. These shoot with power and authority and will take a bit of time and money to construct properly. But there are many others that are simple, cheap, and can be completed quite quickly.

The projects range from a simple slingshot to an elegantly constructed replica Civil War cannon. All of the projects are squarely within the realm of reasonable and interesting science and history, and you'll find nothing here that's malevolent or mean-spirited. This is a book for people like you: responsible people who want to make interesting stuff while having fun and learning something new.

So what will we make? A big potato cannon is here, but it's probably unlike any you might have seen before, even in my first book, *Backyard Ballistics*. This one is more sophisticated, more capable, and better than ever. Imagine making a cannon that can shoot a spud across three football fields, and do so with pretty amazing accuracy.

We'll build a water cannon of tremendous range and unlimited capacity. It's a bigger and better squirt gun than you can buy at any store. Use a five-gallon bucket as your water supply and you can squirt nonstop for a good long while. Mount the water cannon aboard a canoe or small boat and you'll rule the waves.

You want to build a full-auto machine gun? You will find plans and directions for a weapon that can be configured as a single-shot gun designed for a marksman or a fearsome fully automatic weapon. Well, let's not go crazy. It's not really too fearsome—this weapon shoots only marshmallows.

We'll also explore the world of fluid power, using air power to build a Wiffle ball pitching machine that can deliver strike after strike for hours of batting practice. We'll also use air to power a Ping-Pong ball gun that can deliver unhittable serves. And these projects are just a sample of what you can build if you learn a bit of science and history and apply a little basic DIY knowledge.

STUFF YOU'LL NEED

You can find most of the materials that you will need for the projects in hardware and home stores. But in a few cases, you will need a special part that isn't easily found locally. In those instances I've included information on where you can purchase the parts you'll need online.

Besides the materials, you'll need tools. Most of the tools are pretty straightforward and there's a good chance you already own them—electric drills, saws, hammers, and so on. Some of the tools are grouped together for convenience. These include:

- Drill Assortment: A standard assortment of drills includes diameters of $1/16"$, $5/64"$, $3/32"$, $7/64"$, $1/8"$, $9/64"$, $5/32"$, $11/64"$, $3/16"$, $1/4"$. The drills are usually made from a metal alloy called high-speed steel.
- All-Purpose Measuring Tools: You will need to measure materials accurately in almost all of the projects in this book. So there are four important tools I'll group together here. They are:

- A tape measure
- A T-bevel or bevel gauge, which is an adjustable gauge for setting and transferring angles
- A combination square, which consists of a ruled blade and a movable head and which is used to lay out or check right and 45-degree angles
- A grease pen, pencil, or other marker

STAYING IN ONE PIECE

This book is a combination of science, history, and hands-on DIY (do-it-yourself). The science and history make for interesting reading. And, in my opinion, the hands-on stuff is really fun. But before you start building, take a minute to think about what it's going to be like when you actually construct these projects.

Now, it's important to remember that things that shoot equal things that can hurt you. So our first priority is doing everything as safely as possible. I wish I could say we lived in a world where making things and shooting things is a completely safe activity, but we do not. Nonetheless, if you follow the instructions and use your head, odds are you'll not only stay safe, but you'll also be smarter for the experience.

If you want to build the projects that follow, then there are several important points that you need to understand right out of the chute.

The first thing is that some of these projects may not be legal in your area. That may seem unfair, considering that in some places it is perfectly legal to walk down Main Street holding a 12-gauge shotgun in plain sight, but toting a potato gun or a coffee creamer mortar could get you in trouble. But with that said, there appears to be no federal law that regulates the construction and use of the projects contained here, but there may be state, provincial, prefecture, county, shire, or other local laws that do. So check with your local law enforcement regarding the rules that pertain to your area, and then obey them.

Second, recreational artillery projects can by their nature be dangerous. They shoot things, often with great power and force. So treat the items you build with respect. A few important points to keep in mind:

1. Some of the projects that follow shoot with enough force to cause injury. For some of these guns, the expected range can exceed 200 yards. Always use extreme care when handling and aiming the device, and never aim a shooter at people or other creatures. Always wear eye protection.

2. Don't operate a damaged or worn device. Check the items you build frequently for signs of wear. Don't take construction shortcuts.

3. All aspects of all projects must be undertaken or closely supervised by adults.

4. When using tools, chemicals, or other supplies, read and follow all label and instruction book directions.

5. Many of the projects involve the use of plastics. Plastics can become brittle in cold weather and lose strength in hot weather. Be aware of how temperatures can affect your work.

   While not perfect, I believe plastics are an acceptable raw material for many types of recreational artillery projects. Others disagree, but my experience so far has been good.

6. Be aware that the vapors from many chemicals, including PVC plastic pipe cement, are flammable. Allow all joints to fully dry before exposing the device to ignition sources.

7. Use the utmost caution should a dud or misfire occur. Leave it alone until you're certain it's not going to fire. Never look down the barrel, and never point the barrel at anything you don't want to hit.

8. Some projects have specific safety instructions beyond these general ones. Read, understand, and follow those as well.

9. These are perhaps the two most important rules of all and bear repeating:
   Never, ever look down the barrel of any recreational artillery piece, no matter if it's loaded or unloaded, charged or not charged.
   Never, ever aim at something you don't want to shoot.

10. Always remember, things that shoot are things that are dangerous. You'll need to keep your wits about you as you go forward. I can't, and won't, guarantee that by making and using the projects in the book, you won't get hurt. That's just the way things are—materials and tools can fail, you might misinterpret the instructions, you might just encounter some really bad luck, or even (although I try very hard to make sure this doesn't happen) the instructions might not cover everything. Here's the bottom line to all of this: If you go forward with any of the projects in this book, you do so at your own risk. I make no promises as to your safety, and neither does anybody else.

I've made the same statement for all the other books I've written. So far, nobody has come forward to tell me they've had a bad experience with my instructions. As far as I know, nobody's house has burned down, nobody shot themselves with their own spud gun, and nobody wound up doing two to five in Leavenworth. But I guess there's always a chance. So be smart and don't be the first one!

If you don't agree with this, then don't build anything. You can still get a lot out of what's in the pages ahead even if you don't build anything, as the history portion is interesting and learning the science behind legendary and contemporary shooters is worthwhile.

LET'S GET STARTED

So now we get to the good stuff, which is making stuff that shoots. Now, there are a lot of ways to present this information, and I think the best way is to combine DIY instructions, history, and a little science.

Each chapter will tell you how to make a simplified version of some important gun, cannon, or other shooter. Since the first ancient artillerists brought forth what was basically a monster-sized bow and arrow 2,500 years ago, there have been a lot of famous cannons. There's no better way to really understand something than to get some historical background, understand the science behind it, and then cement what you know by getting your hands dirty and building the thing yourself.

With all that said, let's get started with a project that's easy to construct, has a great historical backstory, and above all, provides an extremely satisfying shooting experience.

THE BEVERAGE BOTTLE BAZOOKA

The Beverage Bottle Bazooka is an easy-to-complete project and it's a real thrill to use. As you build it, you will combine a tad of physics with a smidgen of chemistry to make something quite wonderful. And there's a bit of ecology in there as well because the project uses some recycled materials such as cardboard tubes and empty plastic beverage containers. (It's a good way to repurpose and recycle stuff around your house as well.)

HOW THE BAZOOKA GOT ITS NAME

Back in the 1930s, Bob Burns, an American comedian and singer, built a trombone-like instrument out of two pieces of telescoping brass tubing with a whiskey funnel stuck on one end. Burns named his device a bazooka.

When played, the bazooka produces a weird warbling that sounds like a cross between a kazoo and a trombone. Since Burns was such a popular radio entertainer, his contraption also became popular, at least as a novelty, and eventually so did the word *bazooka*. When Burns first coined the phrase, it was taken to mean any strange, stove-pipe-shaped gizmo.

In the early years of World War II, a group of army scientists were looking for a name for a new type of rocket cannon that they had just invented. One of the engineers in the group suddenly had an idea. He looked over the odd-looking gun carefully and then called out that it looked "just like Bob Burns's bazooka." The name stuck.

The new weapon was quickly developed and issued to Allied army forces fighting the German army in Europe. Initially the device had a few bugs, but it was quickly improved through battle testing. Eventually the bazooka became an important part of the army's arsenal.

This soldier should be wearing safety glasses!

MATERIALS

- General purpose glue or hot glue gun
- (1) 3-inch inside diameter (ID)* smooth-to-female pipe thread PVC adapter
- (1) Thick-walled cardboard tube, 3-inch ID, 4 feet long
- (1) 3-inch male pipe threaded plug
- (1) ½-inch PVC pipe, 5 inches long
- 18-inch length of two-conductor 18-gauge speaker wire
- (1) Two-hole rubber stopper, small enough to fit inside the beverage bottle
- Electrical tape
- (1) Piezoelectric gas grill replacement igniter[†]
- (1) 1-inch square wood dowel, 4 inches long
- (1) 2½-inch steel corner brace
- (2) 4-inch diameter hose clamps
- Bottle of rubbing alcohol
- 2- to 2½-inch diameter, 10-inch long empty and clean plastic carbonated beverage bottle[‡]

TOOLS

- Drill and ¾-inch wood drilling bit
- Wire clipper/stripper
- Screwdriver
- Safety glasses

---

* Unless otherwise noted, all pipe diameter specifications in this book refer to inside diameter, which is the way pipes are identified at hardware stores.

† Find piezoelectric grill igniters at hardware stores and other places that sell gear for outdoor barbequing. Online, search for "piezo grill igniter" or "weber 7510." Make sure the igniter you purchase has two visible electrodes: one on the tip and one on the side. If you can't see two, find a different igniter. Generally speaking, piezo grill igniters cost around $10 to $15.

‡ You have quite a bit of latitude when selecting your beverage bottle, but it must be at least ½ inch narrower in diameter than the inside diameter of the cardboard tube. When the propellant in the bottle is ignited, the bottle will expand slightly. If the fit between bottle and tube is too close, the bottle will become stuck inside the tube. I've had excellent luck using the 2-inch diameter, 10-inch long bottles that typically hold fruit-flavored carbonated beverages.

BUILD THE BEVERAGE BOTTLE BAZOOKA

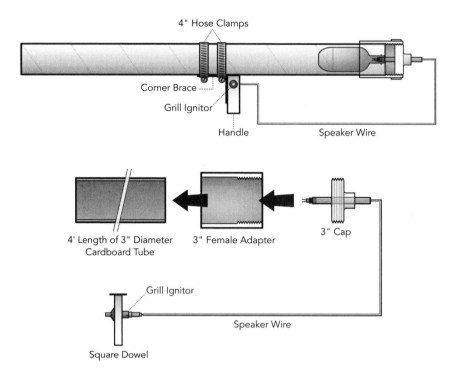

1.1 Beverage Bottle Bazooka Assembly

1. Apply glue to the interior of the 3-inch diameter threaded PVC adapter. Push the adapter onto the end of the cardboard tube. Allow the glue to dry.

2. Drill a ⅞-inch diameter hole in the center of the flat part of the threaded plug. Insert the ½-inch PVC pipe through the hole until it extends approximately halfway in. Glue the pipe into place using glue or hot glue and let dry.

3. Strip off ½ inch of insulation from both ends of both conductors of the speaker wire. Insert the speaker wire into the ½-inch pipe until it exits the other end.

4. Pull apart the conductors for two inches from both sides of the speaker wire. Insert one conductor through each hole in the rubber stopper. Then push the rubber stopper firmly into ½-inch pipe as shown in **diagram 1.1**. Use electrical tape to keep the exposed copper wire in the conductors extended from the end of the stopper and separated by about ³⁄₁₆ of an inch.

1.2 Igniter Detail

5. Drill a hole slightly larger than the diameter of the body of your piezoelectric igniter in the middle of the square wooden dowel as shown in **diagram 1.1**. (This is typically about ¾ inches, but check before drilling.) Insert the igniter into the hole and glue.

1.3 Trigger Detail

Use two of the screws that came with the steel corner brace to attach the brace to the square dowel. Check **diagram 1.1** for the correct orientation.

6. Most replacement piezoelectric igniters come with a wire connector harness. One end of the wire connector harness connects to the electrodes and the other end connects to the spark maker on the grill. For our purposes, we do not need that grill hardware. So simply insert the wires into the appropriate electrode connectors on the piezo igniter and then use a wire

cutter to cut off the unneeded grill spark-making hardware. You can discard the unneeded hardware.

7. Make wire-to-wire connections between the speaker wires' exposed conductors trailing from the ½-inch PVC pipe as shown in **diagram 1.4**. Wrap the connections with electrical tape.

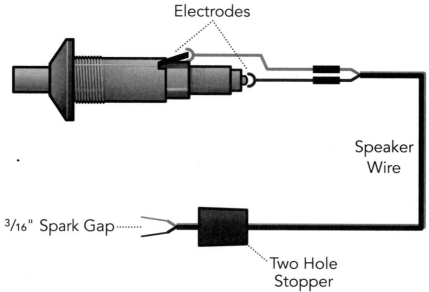

1.4 Igniter to Spark Gap Connection

Test the device by pressing the button on the piezoelectric igniter. If you've done everything correctly, you should see a spark jump between the two exposed conductors. If you don't see a spark, check the spark gap for width, review **diagram 1.4**, and rewire as needed.

1.5 Spark Gap Detail

8. Use the hose clamps to attach the piezoelectric trigger to the bazooka body.

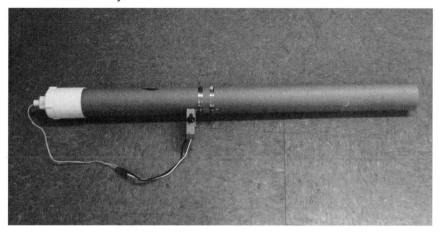

1.6 Beverage Bottle Bazooka

Using Your Beverage Bottle Bazooka Safely

Be aware that your bazooka packs a mighty wallop! While I've never had any trouble, as always, your safety is important, so:

■ Wear safety glasses (and if your ears are sensitive, ear plugs).

■ The bazooka can shoot empty plastic bottles more than 100 feet. Clear the area in front of breakable objects. Never aim the bazooka at anything you don't want to hit.

■ Be careful with rubbing alcohol. It is flammable.

■ Inspect the bottle after every firing and discard when it becomes worn.

■ Always obey local laws and regulations. Check with authorities prior to using if you're unsure about this project's legality in your area.

READY, AIM, FIRE

1. Place a ½ teaspoon of fresh rubbing alcohol in the beverage bottle. Shake the bottle vigorously with your hand covering the mouth of the bottle. Remove your hand and shake out any excess liquid alcohol.
2. Add air back into the bottle by waving it forward through the air four or five times with the bottle mouth uncovered.
3. Quickly insert the bottle onto the ½-inch pipe connected to the threaded plug.
4. Securely screw the plug into the threaded adapter.
5. When ready, press the button on the igniter. Instantly, the alcohol in the bottle will ignite, causing the bottle to rocket forth from the bazooka with amazing energy.

    Enjoy your work!

HOW THE BAZOOKA WAS INVENTED

Prior to World War II, the only way to stop a tank was to blast a hole in it with a large cannon. That meant that armored tanks were pretty much invincible to anything an unarmored infantry platoon could throw at them. But in late 1940, a Swiss engineer named Henri Mohaupt designed a new type of bomb. His invention was an explosive shell that focused the blast into a narrow point of energy capable of penetrating thick armor. It was well adapted to use against armored vehicles because unlike the antitank ammunition shot from cannon barrels, Mohaupt's antitank grenade would work even when it hit a tank at a relatively slow speed.

If a soldier could run up and place this bomb on a tank, it would be possible that for the first time, infantrymen could destroy tanks. This weapon became known as the M10 grenade. Although relatively light, it was still far too heavy for a soldier to toss any distance, and few foot soldiers could get close enough to a German tank to make the attempt.

In 1942, Lt. Edward Uhl, a scientist at the US Army Ordnance Office, was tasked with finding a way to hurl the recently developed M10.

Uhl decided to try attaching the grenade to a rocket and then aiming the rocket at the tank target. The army wanted a grenade launcher as soon as possible, so the development timeline for the project was very short. It didn't take long for Uhl to assemble a prototype by adding propellant, a gas trap, an igniter, and stabilizing fins to an inert grenade. The first test of the rocket grenade took place on a dock on the Potomac River at the army's Aberdeen Proving Grounds in northern Maryland. The results of Uhl's rocket design were exceptionally good. The improvised weapon flew straight and far into the river channel, pretty much where Uhl aimed it.

Now the problem was to design a portable launcher that could hold and aim the M10 and be portable enough for a single man to carry. According to Lieutenant Uhl, the answer occurred to him in a flash of insight.

The bazooka launched a shaped charge rocket.

"I was walking by this scrap pile, and there was a tube that . . . happened to be the same size as the grenade that we were turning into a rocket. I said, 'That's the answer! Put the tube on a soldier's shoulder with the rocket inside, and away it goes.'"

A member of Lieutenant Uhl's team suggested that he attach handgrips to the launch tube to make it easier to handle. Next, he rigged up an electric igniter that sent a charge through a wire to the rocket when the trigger was pulled. At that point, the initial design was more or less complete and all that remained was to test it.

Uhl took the test bazooka back down to the river's edge and put on a welder's mask and gloves. A small group of observers watched. Uhl aimed his rocket out toward the river and squeezed the trigger. Not only did the rocket fly true, but there was little recoil and exhaust to affect the gunner.

Another, more extensive test yielded even better results. With an army general watching closely, Uhl and his team scored a direct hit on a moving tank. People in attendance cheered, and the general asked if he could have a try. After a few minutes' explanation as to how the trigger and sighting worked, the general gave it a try. He also scored a direct hit.

Bazookas have been an important part of the military arsenal ever since.

THE HYDRO SWIVEL GUN

Engineering your own high-performance squirt gun isn't hard if you know some science and are willing to experiment. The science of moving water under pressure is called hydraulics, and the Hydro Swivel Gun is a mixture of basic hydraulic design and the ability to use tools to bring a simple design to reality.

This project can be made in two ways, one a bit more complicated than the other. The first design uses several basic hydraulic components—a cylinder, a piston, two one-way valves or check valves, and a nozzle to produce a high-volume, infinite-capacity water squirter.

The second design eschews the check valves and some plumbing to make a simpler but still great performing squirter. The design of either gun was inspired by the small deck cannons called swivel guns that were used on naval ships in the 1700s and 1800s, the time known as the Age of Sail.

The naval swivel gun, sometimes referred to as a peterero by old-time sailors, was a wickedly powerful but small cannon. It was typically mounted on a universal joint bearing so it could be easily maneuvered into position to shoot in any direction. Lightweight and possessing a high rate of fire compared to heavier weapons, the swivel gun was an important part of the armament of Adm. Horatio Nelson's English navy during the Age of Sail.

Many 18th- and 19th-century naval commanders placed great value upon their swivel guns because they were simple and economical. While it took a crew of 6 to 15 men to operate a big gun like a 24-pounder, a swivel gun was light and easy to load, meaning a couple of gunners could manage it quite easily. And while a swivel was tiny compared to long-range guns, it packed a wallop at close range.

Swivel Gun

By following the instructions on the pages that follow, you can build a high-power, infinite-capacity swivel gun out of easy-to-obtain materials. Unlike those under the command of Lord Nelson, our swivel gun shoots not iron balls but a powerful, high-volume jet of water. This design uses a couple of easy-to-find plumbing fittings called check valves to provide the ability to draw your ammo from any convenient water source, such as a bucket or a lake, on the pull stroke and shoot that water with authority on the push stroke. It's about the most powerful squirt gun you'll ever see!

Making the Hydro Swivel Gun will take anywhere from a few hours to a full day, depending on how handy you are. But once you see how well it shoots, you'll find it's worth the effort!

MATERIALS

- 1¼-inch ID PVC pipe, 2 feet long
- (1) 1⅝-inch ID, ³⁄₁₆-inch wide O-ring
- (2) 1¼-inch PVC caps
- (1) 1¼-inch PVC coupling
- 2-inch ID PVC pipe, 2 feet long
- (2) 2-inch PVC caps
- ¼-inch NPT industrial male pneumatic connector*
- ¾- to ¼-inch NPT reducing bushing* (This may be hard to find, but you can use a ¾ to ½ bushing and another ½ to ¼ bushing instead)
- (2) ¾-inch PVC check valves*
- (3) ¾-inch close pipe nipples*
- ¾-inch PVC tee with female pipe threaded openings*
- (2) ¾-inch NPT nuts* (Check the electrical department of any hardware store; ask for "conduit nuts")
- ¾-inch male NPT to ¾ hose barb connector*
- ¾-inch ID hose, 2½ feet long*
- PVC cement and primer
- Silicone grease
- Pipe sealant*

  * Omit these parts if making the simplified swivel gun. See page 16 for details.

TOOLS

- All-Purpose Measuring Tool Assortment
- Electric drill
- 1⅝-inch drill bit
- Sandpaper
- 1-inch drill bit
- Pipe wrench

BUILD THE HYDRO SWIVEL GUN

The Hydro Swivel Gun is made up of two main sections—the plunger and the squirt-suction head. The plunger is a close-fitting piston that the squirt gunner extends and retracts in order to draw water in and out.

The squirt-suction head controls the flow of the water, allowing water to enter the plunger tube when the plunger is retracted and spray out of the nozzle when the plunger is pushed forward.

1. Begin by examining the detail on **diagram 2.1**. Cut a 2-inch-long segment from the 2-foot-long 1¼-inch diameter PVC pipe. Slide the O-ring over the pipe and center it. Note that the short pipe with the O-ring on it fits between the 1¼-inch cap and coupling. The purpose of the O-ring is to seal against the 2-inch pipe but still allow one pipe to slide easily upon the other.

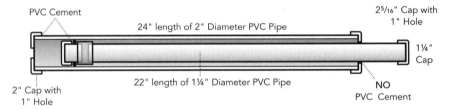

PVC Cement
24" length of 2" Diameter PVC Pipe
2⁵⁄₁₆" Cap with 1" Hole
1¼" Cap
2" Cap with 1" Hole
22" length of 1¼" Diameter PVC Pipe
NO PVC Cement

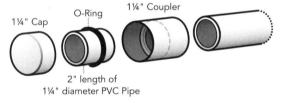

1¼" Cap
O-Ring
1¼" Coupler
2" length of 1¼" diameter PVC Pipe

2.1 Plunger Assembly

2. Drill a 1⅝-inch hole in the center of the one of the 2-inch caps. Use sandpaper to enlarge the hole until the 1¼-inch diameter pipe can slide through with no interference. (The 1¼-inch dimension refers to the inside diameter of the pipe. The pipe has a roughly 1.67-inch outside diameter.)
3. Drill a 1-inch hole in the center of the other 2-inch cap.

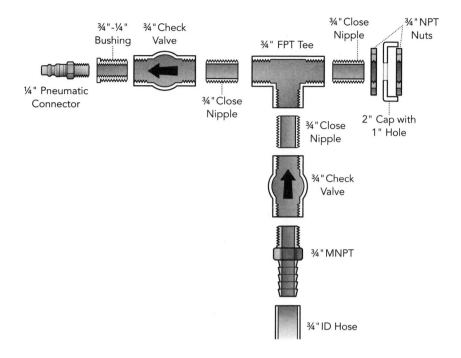

2.2 Squirt-Suction Head Assembly

4. You are ready to assemble the squirt-suction head. Refer
    to **diagram 2.2** to review the details of how these parts fit
    together. Be sure to note the flow direction on the check valves
    (there are arrows molded into the plastic showing this) and
    make sure they agree with the arrows on the diagram.

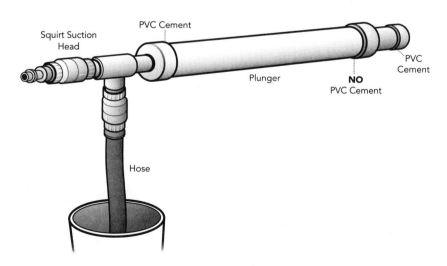

2.3 Hydro Swivel Gun Assembly

5. After completing the squirt-suction head, attach the assembly to the plunger. **Diagram 2.3** shows how to fit the parts together. Use the pipe wrench to fasten a conduit nut on each side of a ¾-inch close nipple to secure the squirt-suction head assembly to the 2-inch end cap with the 1-inch diameter hole, as shown on the right side of **diagram 2.2.**

   You'll need access to the O-ring from time to time in order to grease it, so don't solvent-weld the cap with the larger hole to the pipe. Use PVC cement and primer to seal the other caps to the PVC pipes; **diagrams 2.1** and **2.3** both indicate which caps to seal and the one to leave accessible. (See the appendix for detailed information on how to solvent weld PVC pipe.)

6. Use pipe sealant around the conduit nut connections (as well as any other places you notice leaks).

AVAST, ME HEARTIES!

To operate your swivel gun, lubricate the O-ring with plenty of waterproof grease. Then extend the hose into a bucket or pond. Pump the 1¼-inch pipe back and forth until the gun sprays water.

   Feel free to experiment with other nozzle sizes and shapes until you get the sort of water action you like best. You can suspend the gun from ropes or place it on an improvised stand to turn your hand cannon into a full-fledged swivel gun.

Build a Simplified Swivel Squirt Gun

If the plumbing on the squirt-suction head seems too complex for you to build, fear not! You can build the gun without this section. Just omit all the parts shown on **diagram 2.2**, the Squirt-Suction Head Assembly, and simply drill a ⅜-inch hole (in place of the of the 1-inch hole) in the center of the 2-inch cap. It won't be nearly as fast a squirt weapon, but it will still spray a lot of water!

   To draw water into the simplified gun, just submerge the front end of the gun in a bucket or a pond and retract the piston. To fire, just aim and give the piston rod a hearty push.

THE SCIENCE BEHIND YOUR HYDRO SWIVEL GUN

The Hydro Swivel Gun is what hydraulic engineers would term a "two-valve forcing pump." It's pretty much based on the action of check valves. Check valves are simple little devices that allow fluid to flow in only one direction.

There are different types of check valves, but the one used in the Hydro Swivel Gun is called a ball check valve, and it's probably the simplest to understand.

There are two openings in a ball check valve, one where the water comes into the valve and one where it goes out. In between, there is a ball held in place by a spring.

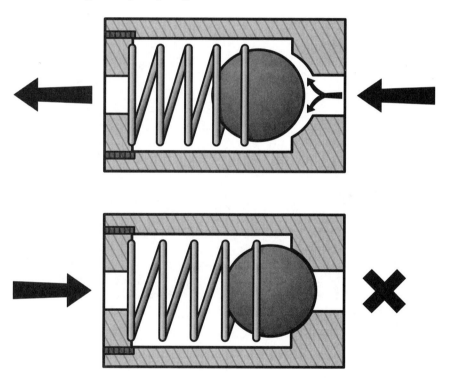

2.4 Check Valve

When the fluid goes from right to left as shown in **diagram 2.4**, the ball is pushed forward. Because the ball is restrained, or checked, by the spring, the fluid can move around the ball and through the valve. But the when the flow moves in the other direction, the spring pushes the ball up against the inlet opening, which the ball seals completely.

In our Hydro Swivel Gun, when the 1¼-inch piston is retracted, a vacuum is created inside the cylinder. Because of the check valve, no air can enter through the nozzle, so the cylinder is evacuated of air. This creates a vacuum inside the gun barrel, and so the gun sucks up water through the hose. When the piston is pushed forward, water shoots through the nozzle check valve because the hose check valve prevents the water from exiting through it.

Admiral Nelson and His Powder Monkeys

In 1805, British admiral Horatio Nelson and his fleet of warships engaged the combined navies of Spain and France. The location for this history-changing event was off the coast of Spain at a place called Trafalgar.

Battle of Trafalgar

Although the Spanish and French had considerably more guns on board their ships, the English won the day. The well-trained English gun crews could fire three times faster than their opponents, making each British cannon a far more formidable weapon. Nelson's victory at Trafalgar was complete and for the next hundred years, Britannia ruled the waves.

British Naval Gun

By the early 19th century, naval gunnery techniques had become quite refined, and the roles of each man on a gun crew were precisely defined. The man in charge of the gun was known as the gun captain or simply, the gunner. He stood to the rear of the gun and had the job of aiming the cannon by inserting a wedge called a quoin under the barrel to elevate or lower the muzzle. He was also the person who fired the gun when the target was in his sights. Next to him was the matross, or gunner's mate, who assisted the gunner. Next in line were the handspike men, who used long spikes to turn and raise the barrel. The gun, being so large and heavy, was difficult to move. So additional men, called tackle men, also jumped in to help pull the gun into firing position.

In front of the handspike men and tackle men stood the loaders, who rammed the cannonballs and powder down the barrel. Beside them crouched the spongers. Once the gun was fired, the spongers used their long-handled sponges to mop out the barrel to make sure there was nothing left burning inside. Once it was clean, the firing process could be repeated.

There was one other member of the crew, the powder boy, also known as the powder monkey. Don't let the name fool you; the job of the powder monkey was as important

Age of Sail Gun Crew

as it was dangerous. Since gunpowder is such a hazardous chemical compound, it was stored deep in the hull of the ship, away from flames and fire, and only enough powder was kept near the gun as was required to fire a single volley. Someone would have to run down to the powder storage area, or magazine, after each shot to get another bag, and that job typically fell to the youngest person on the ship, the powder monkey.

Powder boys were as young as 11 years old. It's hard for modern people to imagine a situation where boys who were too young for the sixth grade would leave their families to spend years among rough-and-tumble sailors and fight in naval battles, but 300 years ago it was common.

Powder Monkey

The Battle of Trafalgar was Admiral Nelson's last and greatest victory. In five hours of intense fighting, the British routed the enemy fleet, destroying 19 ships. Although the British did not lose a ship, it was a hard-won victory as British casualties were heavy; around 1,500 British seamen were killed or wounded, including Admiral Nelson himself.

During the Battle of Trafalgar, the job of the powder monkey was critical. The powder monkeys ran from their assigned gun crew through the passageways and down the hatchways of the ship to the dark, thickly walled magazine where they yelled to powder men working behind heavy water-soaked curtains. When big, sweating hands holding a six-pound bag of powder shoved aside the curtain, the powder monkey grabbed the bag and slung it over his shoulder. Then he rushed back up to the gun deck to deliver the charge.

During an engagement like the Battle of Trafalgar, the stream of powder monkeys rushing between the magazines and the guns was continuous. If one boy could not continue, another one jumped in to take his place. The powder monkey had to be a fast mover indeed.

It was not a job most people would care to have. During a naval engagement, ships were taking fire as frequently as giving it, so cannonballs would smash into the sides of the ships, spraying splintered wood inside. And while it wasn't common, it wasn't unknown for a cannon to explode when it was fired, tearing up the gun crews with shrapnel.

The powder monkeys worked steadily through it all, typically serving with diligence and heroism despite their youth. They were paid little, had no bargaining power, and never dared complain.

THE MACK-MACK GUN

In World War I, Allied combat radio operators developed a code to make sure they would be understood properly when the quality of a phone line was poor or the noise of battle made it difficult to hear. This code, still in use today, is referred to as the phonetic alphabet and using it improves understanding in hard-to-hear situations.

For example, instead of requesting assistance in sector A-3 or J-9, a World War I military radioman would say Ack-Tree or Johnnie-Niner. By substituting easily distinguishable syllables and words like Ack and Johnnie for the single letters A and J, battlefield communications improved.

So when a new cannon for shooting down German aircraft flying over the trenches was unveiled, British soldiers started calling it an Ack-Ack gun, the phonetic code for antiaircraft gun.

Regular soldiers soon caught on to using the new lingo and started to use the new alphabet in their daily lives as well. Over time, the method has become standardized. Most English-speaking military use the NATO phonetic alphabet.

A Alpha	**H** Hotel	**O** Oscar	**V** Victor
B Bravo	**I** India	**P** Papa	**W** Whiskey
C Charlie	**J** Juliet	**Q** Quebec	**X** X-ray
D Delta	**K** Kilo	**R** Romeo	**Y** Yankee
E Echo	**L** Lima	**S** Sierra	**Z** Zulu
F Foxtrot	**M** Mike	**T** Tango	
G Golf	**N** November	**U** Uniform	

The first really effective Ack-Ack guns were the British QF-1 and QF-2 models used during the First World War. These were early types of a revolutionary new type of gun called an autocannon, and it was engineered so that it could fire a whole boxful of projectiles rapidly.

War planners of the early 20th century put a high priority on developing methods to protect ships from attacking airplanes, and they quickly realized that autocannons were right for the job. With such great emphasis placed on autocannons, progress was rapid.

Probably the most successful antiaircraft weapon system ever designed was the 40-millimeter Bofors gun. Designed by the Swedish arms maker AB Bofors in the mid-1930s, these weapons were manufactured by the tens of thousands in the United States and placed on US Navy warships throughout World War II.

Bofors Guns

Bofors guns were complicated, expensive, and effective. The twin-barreled autocannons could paint the sky with lead, firing more than 120 antiaircraft shells per minute. They were extremely successful at protecting ships from enemy planes. For example, on one occasion in 1942, Bofors guns aboard the US battleship *South Dakota* shot down 32 enemy planes in half an hour. While that's hardly typical, it does show how effective fast-firing antiaircraft guns can be.

Let's make our own Ack-Ack gun, which we'll call a Mack-Mack Gun because it's designed to fire marshmallows instead of two-pound lead slugs. Now, before you write this project off as being strictly for the kids because it shoots only mini-marshmallows, take a look at what this thing can do: this project is a high-velocity, breech-loading, compressed-air-powered gun with a full-auto-fire option. Yup, it's definitely a step above most other marshmallow guns.

That said, it's still a marshmallow gun, and I believe it's fine for children of any age to use.

Also, you have considerable latitude to make changes. Feel free to try longer or shorter pieces, extra joints, or different angles.

MATERIALS

All parts are ½-inch diameter, schedule 40 PVC pipe or pipe fittings unless noted otherwise.

- (3) PVC pipe caps
- (1) Air tank valve, ¼-inch NPT to Schrader*
- (1) 16-inch long PVC pipe
- (3) 4-inch long PVC pipes

---

* These valves are available at hardware and home stores with reasonably large inventories. You can also find them online by Internet searching for "air tank valve" or "tru-flate valve." Another alternative is to order from a large industrial supply company like McMaster-Carr.

- (3) 2-inch long PVC pipes
- (1) PVC coupling fitting
- (2) PVC tee fittings
- (2) PVC 45-degree elbow fittings
- (1) CO_2-powered bicycle tire inflator with several spare 16 gram CO_2 cartridges*
- (1) Bag mini-marshmallows

TOOLS

- All-Purpose Measuring Tool Assortment
- Electric drill
- $7/32$-inch bit
- ¼-inch pipe tap and handle[†]
- Adjustable wrench
- Safety glasses

Optional: rotary tool (such as a Dremel) with a thin cutting wheel attachment and rotary sanding drum[‡]

BUILD THE MACK-MACK GUN

1. You'll need to make a threaded hole in one of the pipe caps. First drill a $7/32$-inch hole in the center of one of the pipe caps. Now you'll need to cut screw threads into that hole. This process is called tapping a hole.

   The tool you use to cut a male threaded piece is called a die. When you make a female threaded piece, you use a tap. In this case, we are making threads in a hole, which means you're making a female threaded piece. So you need a tap. Here's how to thread a hole:

   a. Start the threading process by carefully positioning the axis of the tap perpendicular to the hole you just drilled.

---

* Try finding these at a bike store or search online for "CO2 bike pump" or "CO2 bike tire inflator."

† A ¼-inch pipe tap is different than a ¼-inch UNC tap. I know that's confusing, but that's the way things are.

‡ This is for optional quick-breech-loading ability. You can assemble the Mack-Mack without it and still have a working, although slower, single-shot weapon.

b. Turn the tap a half turn and then back out the tap a quarter turn to remove plastic shavings so the tap doesn't get clogged.

c. Keep doing this until the hole is fully threaded. The plastic cuts easily, so you shouldn't have much trouble with this.

2. Insert the tank valve into the hole and tighten using the adjustable wrench.

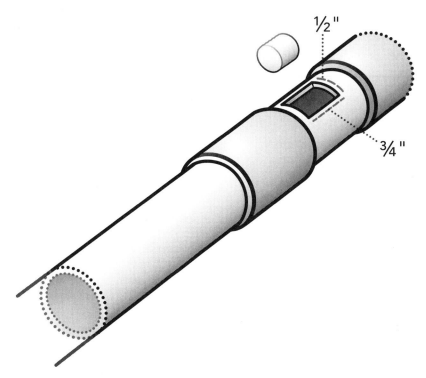

3.1 Quick-Breech-Loading Detail

3. Optional: For quick-breech-loading capability, cut a ½-inch by ¾-inch hole in the 16-inch-long barrel piece using the rotary tool with a thin cutting blade inserted in its chuck. Cut the hole approximately 1 inch from one end. See **diagram 3.1** for details on where to make the cuts.

Then switch to the sanding drum and enlarge the interior of the plastic coupling just until the coupling slides easily over the pipe.

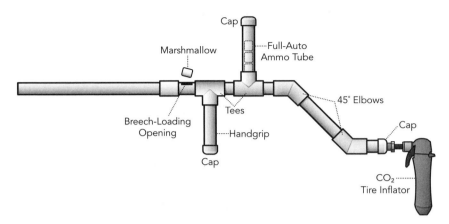

3.2 Mack-Mack Assembly

4. Assemble the Mack-Mack as shown in **diagram 3.2**. Press fit the pipe pieces into the sockets of the fittings, but do not use PVC cement or primer.
5. Screw the CO_2 bike tire inflator into the Schrader side of the air tank valve.

SHOOTING YOUR MACK-MACK GUN

Full-Auto Burst Fire Mode

In the world of automatic weaponry, burst mode or burst fire mode means you can fire a predetermined number of projectiles with a single pull of the trigger.

1. Remove the cap from the upward-pointing 4-inch pipe.
2. Insert six or seven mini-marshmallows into the uncapped pipe, making sure they can fall freely into the main gun barrel. Replace the cap.
3. Put on safety glasses and choose a target wisely. To shoot, pull the trigger on the bicycle tire inflator and release quickly. The mini-marshmallows will shoot out, one after another, in a glorious fusillade of soft, sugary firepower. Go easy on the trigger—pulling on it for too long will quickly deplete your CO_2 supply and cause the cartridge to ice over. If that happens, you'll need to wait until it thaws before you can fire again. Expect to get about six to eight bursts per gas cartridge.

Single-Shot, Quick-Loading Sniper Mode

1. Push the sliding coupling on the barrel forward to expose the loading hole in the barrel.
2. Insert a mini-marshmallow. (All of the mini-marshmallow brands I found fit perfectly in the barrel. But if you have trouble, reduce the size of the marshmallow a bit by cutting off small pieces until it fits, or buy a different brand.)

   Pull the coupling back into place so it covers up the breech. The coupling should slide smoothly over the barrel but still fit snugly.
3. Choose your Mack-Mack target wisely. To shoot, pull the trigger on the bicycle tire inflator and release quickly. The marshmallow will shoot out at high speed. Keep a light touch on the trigger. If you do it correctly, you will get 10 to 12 shots per cartridge.

Pro Tip: You can use your lungs instead of the cartridge if you want. Just remove the tire inflator and the end cap with the tank valve and simply blow hard into the gun to fire the marshmallows.

THE SCIENCE INSIDE: CARBON DIOXIDE CARTRIDGES

Have you ever wondered why the small steel cartridges of compressed gas that power the Mack-Mack (and regular air rifles as well) are filled with carbon dioxide instead of compressed air? It's because carbon dioxide is a much different animal than plain old air. In many ways, CO_2 (the chemical abbreviation for carbon dioxide) behaves in a much more complicated fashion.

If you measure the pressure inside a 16-gram CO_2 cartridge, you would find there's a lot of it in there: around 850 psi at normal room temperature. That's because CO_2 has an enormously high vapor pressure. That means that if there's any liquid inside the cartridge at all, the gas pressure in whatever space exists above it will be around 850 psi at temperatures around 70 degrees F. No wonder the cartridges are made out of heavy steel!

When you push the Mack-Mack trigger, some of the gas is released and the liquid remaining inside the cartridge instantaneously turns to gas until the pressure goes back up to that superhigh vapor pressure. The interesting thing to note is that unlike a pressurized air tank, the pressure in a CO_2 tank is determined by temperature, not by mechanical compression. This means that the CO_2-powered Mack-Mack will shoot with the same power and velocity until all the liquid inside the cartridge is gone.

But there's even more science inside that metal cartridge to consider. When you shoot, you'll notice that the cartridge becomes noticeably colder every time you press the trigger. Why is that?

When you press the trigger and the CO_2 exits the cartridge, only the gas shoots out, leaving liquid behind. Since there is now more liquid than gas inside the cartridge, the pressure inside temporarily drops.

There is a law of physics called Gay-Lussac's Law that states that pressure and temperature are directly proportional. So when the pressure drops in a closed container, the temperature also drops. That's why CO_2 cartridges can get so cold after use.

But the cartridge is surrounded by much warmer air and immediately begins to warm back up. As it does, the heat causes some of the liquid CO_2 inside to boil. The balance between liquid and gas returns to how it was before the trigger was pulled, and the pressure inside goes back up. This continues until all of the liquid CO_2 is used up. But because of the pressure inside the cartridge, if you press the trigger too quickly and too often, the Mack-Mack could ice up and the pressure will drop precipitously until it warms back up.

Safety Tip: Follow the CO_2 cartridge manufacturer's safety and handling instructions carefully! Don't heat or puncture the cartridge or try to refill them. Also, be aware that the cartridge can become very cold during use.

THE LITTLE CORPORAL CANNON

It takes a ton of scientific knowledge to design and operate a cannon. In fact, there's so much here that this scientific specialty has been given its own name—ballistics—and it's an important subset of classical physics. But it's even bigger than that; so big, in fact, that the field of ballistics has been subdivided into smaller, more manageable sections. For example, if you're studying what's happening inside a cannon barrel when the powder is ignited, then you're studying the field of interior ballistics. If you are looking at what the projectile does after it exits the cannon, that's exterior ballistics. Finally, if your area of investigation is what the projectile does to the thing it hits, that's called target ballistics.

Yes, that's a lot to know, but if a cannoneer understands all of that, he or she might become the most important person on the battlefield. Consider for a moment the nickname that cannons and other pieces of artillery have been given by soldiers. They're often called the King of Battle because cannons, both big and small, make such an impact on the battlefield.

While artillery has played a role in nearly every conflict since the 15th century, it played a particularly important role in the American Civil War. The most widely used, dependable, and deadly cannon of the Civil War was the Model 1857 smoothbore howitzer. This 12-pounder (meaning it could shoot a 12-pound cannonball) was effective, reliable, and easily transported.

Although its official name was the Light 12-Pounder Field Gun, it was better known by its nickname: the Napoleon.

The project in this chapter is called the Little Corporal Cannon because it is inspired by the 1857 Napoleon, and one of the nicknames given to Napoléon Bonaparte was "the little corporal." You'll find this project to be perhaps the biggest, loudest, and most impressive in this book. When it fires, your neighbors are sure to know! And I've got some good news for you: there's nothing particularly tricky or hard to figure out about building it.

The Little Corporal described below is not designed to shoot projectiles. Rather, it's a special sort of cannon known as a salute cannon, designed to make very loud gestures of respect. (You've probably heard the phrase "21 gun salute," right?)

The cannon parts themselves are not expensive, although if you outfit it with large, store-bought wheels as shown in the photograph, that could run into some money. You can save some dough by cutting your own wheels out of plywood or building a wooden gun carriage of your own design.

**The Napoleon
12-Pounder**

MATERIALS

- (3) 1-foot pieces of 2 × 10 lumber
- (1) 5-foot piece of 2 × 6 lumber
- (8) Deck screws, 2½ inches long
- (2) ⅜-inch bolts, 4 inches long, washer, wingnut
- (2) Medium cabinet door handles
- (2) Medium rope cleats
- (1) ¾-inch threaded pipe, 24 inches long
- (2) ¾-inch pipe straps
- 30-inch wheels (Do an Internet search on "30 inch wooden wheels" or cut your own from plywood)
- (2) ¾-inch pipe caps
- PVC primer and cement
- Heavy-duty strapping tape
- Cotton cloth, straight wood stick, and charcoal lighter fluid for linstock
- Calcium carbide or Bangsite*
- Matches

All of the parts below are schedule 40 PVC pipe.

- (3) 2-inch socket end caps
- (1) 2-inch diameter pipe, 4 inches long
- (2) 2-inch crosses
- (2) 2-inch diameter pipes, 3 inches long
- (2) 2-inch socket to female NPT adapter
- (2) 2-inch diameter NPT threaded end plugs
- (1) 2-inch diameter pipe, 30 inches long
- (2) 2-inch diameter pipe, 6 inches long
- (1) 2- to 2½-inch PVC reducing fitting

---

* Many vendors sell small quantities of calcium carbide online. It's used for a variety of purposes, including powering the headlamps that coal miners and spelunkers use. Conduct an Internet search using the key words "calcium carbide" to find companies that sell it. A particularly convenient product is called Bangsite and you can buy it at www.bigbangcannons.com. Bangsite is often available at local hobby stores as well.

TOOLS REQUIRED

- All-Purpose Measuring Tool Assortment
- Hand saw or miter saw, and 2½-inch hole saw
- Electric drill
- Twist Drill Assortment
- Digital or analog scale
- Earplugs or ear protection

BUILD THE LITTLE CORPORAL CANNON

1. Cut your wood pieces to size using a hand or miter saw. If you are cutting your own wheels from a plywood sheet, tap a nail into the center of the wood panel you plan to use. Tie one end of a 15-inch piece of strong string or cord to the nail and the other end to the shoe of a jigsaw. If you can keep the cord taut as you cut, you'll have a (nearly) perfect circle when you're finished.

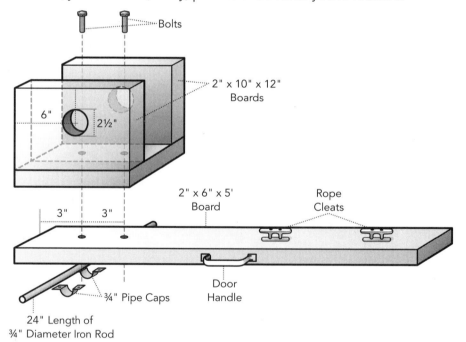

4.1 Little Corporal Cheeks and Stock Trail

2. **Diagram 4.1** shows the location of the holes you will need to drill. Drill a 2½-inch hole in the center of two of the 1-foot long

2 × 10 pieces of lumber. Drill ⁷⁄₁₆-inch holes in the center of the remaining 1-foot-long 2 × 10 piece of lumber at 3 and 6 inches from one end.

3. Using the measuring tools, identify and mark the ⁷⁄₁₆-inch bolt hole locations that connect the cheeks to the 5-foot-long 2 × 6 trail stock as shown in **diagram 4.1**.

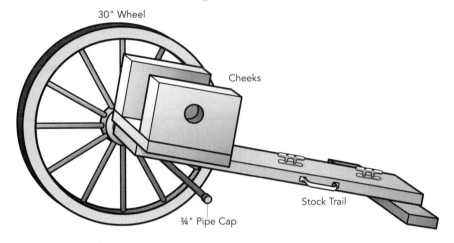

30" Wheel

Cheeks

Stock Trail

¾" Pipe Cap

4.2 Carriage Assembly

4. Assemble the cannon cheeks as shown in **diagram 4.1** using deck screws.
5. Attach the cannon cheeks to the trail stock using the ⅜-inch diameter bolts, washers, and wing nuts.
6. Attach the handles (artillerymen would call these "dolphins") and the cleats (which were known as "prolonge hooks") to the trail stock with screws as shown in **diagrams 4.1** and **4.2**.
7. Attach the ¾-inch pipe axle to the trail stock using the pipe straps as shown.
8. Slide the wheels onto the pipe axle. Attach the wheels to the axle using the pipe caps to hold them onto the axle.

   The carriage is complete. Now it's time to make the gun. While this isn't particularly difficult, it might take you some time to figure out which part is which, especially if you've never worked with PVC pipe before. It's a good idea to review the section in the appendix that explains the different types of PVC fittings (see page 124).

Fit all the pieces together prior to solvent welding to make sure you understand how it all goes together. Once you start gluing stuff together, there's no fixing mistakes!

There are two main parts to the gun, the breech and the barrel. The short length of 2-inch diameter pipe connects the two main parts together.

BUILD THE BREECH

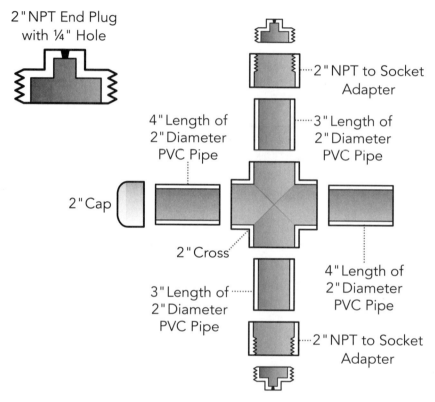

2" NPT End Plug with ¼" Hole

2" NPT to Socket Adapter

4" Length of 2" Diameter PVC Pipe

3" Length of 2" Diameter PVC Pipe

2" Cap

2" Cross

3" Length of 2" Diameter PVC Pipe

4" Length of 2" Diameter PVC Pipe

2" NPT to Socket Adapter

4.3 Breech Assembly

1. Begin assembling the breech by referring to **diagram 4.3**. Assemble the breech by solvent welding one of 2-inch socket caps to the 4-inch long, 2-inch diameter pipe. Solvent weld the pipe into one opening on one of the crosses.
2. Insert the two 3-inch-long, 2-inch diameter pipes into two opposite openings of the 2-inch cross. Solvent weld a 2-inch diameter threaded adapter to each pipe.

3. Solvent weld the 4-inch long, 2-inch diameter pipe to the remaining opening in the 2-inch cross.

4. Drill a ¼-inch hole in one of the 2-inch threaded plugs. Screw on the threaded plugs. The plug with the hole is the top.

The breech is now complete.

BUILD THE BARREL

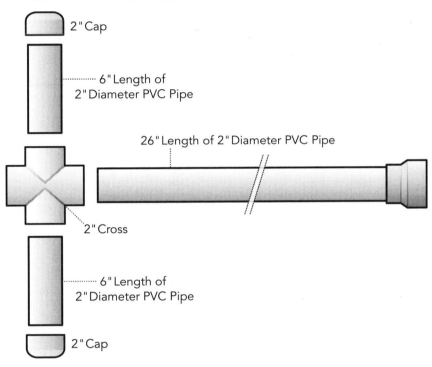

4.4 Barrel Assembly

1. Begin assembling the barrel by referring to **diagram 4.4**. Solvent weld the 30-inch pipe to one end of the 2-inch diameter cross. You can make the barrel a bit longer or shorter to suit your wishes. A shorter barrel makes for a louder gun.

2. Place the barrel assembly on the carriage with the openings of the 2-inch cross aligned with the holes on the wooden cheeks. Insert the two 6-inch long, 2-inch diameter pipes through the holes in the wooden cheeks and into the openings on the cross. Solvent weld the pipes to the cross and then solvent weld the 2-inch end caps to the 2-inch diameter pipes as shown in **diagram 4.4**.

3. Place the 2-inch to 2½-inch reducing fitting on the end of the barrel. (This piece is purely cosmetic and may be omitted if you have trouble finding this fitting.)

FINAL ASSEMBLY OF THE CANNON

1. Align the holes in the two crosses so they are perpendicular to one another, as shown in **4.5**. Solvent weld the 2 inch pipe extending from the first cross to the opening on the second cross.

4.5 Little Corporal Rear View

2. Wrap the barrel with heavy-duty strapping tape.

WAKING UP THE NEIGHBORS

4.6 Little Corporal Cannon

1. Make a linstock, which is a stick with a fork at one end that holds a burning fuse called slow match. You don't want to hold a match directly over the touch hole because you could get burned. Instead of slow match, you can use a piece of cloth or a cotton ball. Make your linstock about a foot long.

   To make a simple linstock, wrap a bit of cotton cloth around the end of a straight wood stick and add a few drops of charcoal lighter fluid to the piece of cotton.

2. Clear the area in front of the cannon of people and animals. Unscrew the top 2-inch threaded end plug. Pour a couple ounces of water into the reservoir formed by the bottom plug. Don't raise the cannon more than a few degrees because the water inside will spill out of the reservoir.

3. Use a scale to weigh out no more than half a gram (see safety notes on page 40) of calcium carbide (Bangsite powder is a widely available form of pulverized calcium carbide). Pour the calcium carbide into the water in the reservoir. The calcium carbide will form acetylene gas when it dissolves in water. Screw the top plug finger tight, making sure it is secure and cannot blow off.

4. Ignite the cotton end of the linstock and bring the flame to the tapered hole in the top end plug, keeping your hand away from the firing hole. The cannon will fire with a very loud bang!

5. Carbon dioxide and other vapors may linger in the barrel after firing. You will need to blow the old gas out or simply wait for it to clear prior to refiring the cannon.

6. If you have trouble unscrewing the plugs, use a wrench.

Safety Notes

- Acetylene gas is formed when calcium carbide is mixed into water. It is very powerful and you must respect that. A half gram of calcium carbide is plenty to start. Use your own judgment when determining the maximum amount to use.
- Do not put any projectile inside the cannon barrel. This project is for making noise; it is not designed for shooting anything.
- Always use a linstock or long-handled lighter to keep your hand away from the touch hole when you fire the cannon.
- The cannon is very, very loud. Protect your ears accordingly.
- Never look down the barrel.
- Keep the area in front of the cannon clear of people, breakable things, and animals.
- Blow air into the breech after firing to remove the products of combustion. The cannon needs fresh air to fire.
- Wash out the cannon breech when it looks dirty.

THE PHYSICS OF RIFLING

The Little Corporal's inspiration was the 1857 Napoleon cannon, which was named after Prince Charles Louis-Napoléon Bonaparte, nephew of the French emperor Napoléon Bonaparte. The 1857 artillery piece was a smoothbore, meaning that the cannon barrel was a simple smooth tube, without the spiraled snake of metal called rifling that the more advanced cannons of the period were using.

Rifled cannon barrels cause projectiles to spin as they hurtle through the air, and they are unquestionably more precise than smoothbore. Still, the 1857 Napoleon was reasonably accurate and robustly built. Its barrel, free of the obstructions of rifling, was much quicker to clean and reload during battle than nearly any other gun. For that reason, it became the favorite field gun of both the Union and Confederate armies in the American Civil War.

Rifling causes oblong projectiles like cannon shells to spin as they travel down the barrel of the gun. The spinning causes the projectile to increase a property called angular momentum. To understand what angular momentum is, first consider regular, linear momentum. Linear momentum is the tendency of a thing moving in a straight line to continue moving in a straight line unless something acts on the thing to stop that.

So too does angular momentum work—a thing in rotation will continue to rotate because of its angular momentum unless something happens to slow it down and transfer that angular momentum to something else. This basic law of physics was first described by Isaac Newton. As he put it, momentum is always conserved, and that's true for objects moving in straight lines or spinning around an axis.

Just as things moving forward in a straight line keep on doing so unless they are acted upon by an outside force, so do spinning bullets and cannon shells resist tumbling end over end or spinning in any new direction around any new axis as they fly through the air. This is the reason—the law of momentum conservation—that they travel farther and more accurately.

Artillery Hell

The 1857 Napoleon played key roles in all major Civil War battles but probably none so much as on the deadliest single-day battle: the Battle of Antietam.

In mid-September 1862, Confederate general Robert E. Lee received news that his subordinates had captured the strategic town of Harpers Ferry, West Virginia. Adding this news to his knowledge that the Union general George McClellan was habitually slow to act and cautious when he did so, Lee decided to engage the larger Union Army at Sharpsville, Maryland. While he had a force about half the size of McClellan's, Lee knew reinforcements were on the way, and he believed that if he was bold and lucky enough to be successful, a victory here might be an opportunity to turn the war decisively in the South's favor.

On September 17, Union commanders launched a series of attacks on Confederate battle lines at Sharpsville. The result was thousands of casualties on both sides. Men were cut to pieces by rifle fire, but most horrible were

The Battle of Antietam

the casualties attributed to artillery. Hundreds of cannon, including 130 Union and 30 Confederate Napoleons, inflicted so many casualties that the day has gone down in history as "Artillery Hell." The exact number of men killed is not precisely known, but the estimates are about 24,000 men cut down, about equally split between blue and gray.

The youngest person manning the cannons on the battlefield of Antietam was a teenager from Ohio named Johnny Cook. When the Civil War broke out in 1861, thousands of men rushed to their local enlistment offices to join up. In June, so did Johnny; he was given a uniform and a bugle and assigned to the Battery B of the 4th US Artillery. But unlike most other members of Battery B, Johnny had not yet reached his 14th birthday!

Battery B consisted of 150 officers and enlisted men armed with six bronze-barreled Napoleon cannons that fired 12-pound projectiles. The unit spent the first year of the war defending Washington, DC, and then became engaged in a series of actions that ranged from minor skirmishes to major battles.

In September 1862, the 4th Artillery was assigned to General George McClellan's Army of the Potomac. The unit marched along until they reached the banks of Antietam Creek. At this point, the now 15-year-old Johnny Cook was assigned as an aide to an artillery officer named James Stewart.

On the morning of September 17, Stewart, with young bugler Cook at his side, was in charge of two Union cannons.

Cook wrote later that no sooner had Stewart's men set up the guns "when a column of Confederate infantry, emerging from the so-called west woods, poured a volley into us, which brought 14 or 17 of my brave comrades to the ground."

Normally, it took a crew of five trained men to fire a Napoleon. One man cleaned and greased the gun; another loaded in the shell and powder; a third primed the gun; the chief gunner sighted for distance and trajectory. And a fifth man pulled the lanyard at the command to fire.

At his station on the front line, Cook saw that two of the battery's guns were silent. Dead and wounded soldiers lay nearby. Cook made his way to the cannon battery. Taking a pouch of powder from a dead cannoneer, he manned the cannon. By dint of his immense effort amplified by what must have been intense fear, the boy worked the entire gun all by himself.

"We were then in the vortex of the battle," he wrote later. "The enemy had made three desperate attempts to capture us, the last time coming within 10 or 15 feet of our guns."

Soon another soldier rushed up to help. That soldier was none other than General John Gibbon, the brigade commander. With Cook and Gibbon firing upon the rebel soldiers, Battery B managed to fend off the Confederate charge, and while doing that, wrote a unique page in the annals of military history.

Strange things happen during wartime. It is highly likely there had never before been and will never again be such an unusual pair of fighters, a 15-year-old bugler working a cannon with the aid of a brigadier general.

Cook saw many more battles as a member of Battery B, including Gettysburg. There, Stewart, who had been promoted to captain, assigned Cook to carry messages through the battlefield. Thirty-two years later, Johnny Cook was awarded the army's highest award, the Medal of Honor, the youngest person ever to win one.

THE TEMPERED STEEL RUBBER BAND SHOOTER

R ubber band guns (often known simply as RBGs) are an easy way to get your feet wet in the world of suburban siegecraft. Rubber bands are cheap and not particularly dangerous (still, wear safety glasses because you don't want to take one in your eye), yet there's a great deal of creativity possible as you consider the best construction options for your rubber-powered cannon. While most RBGs are made of wood, here's one that's a step up—one made of heat-treated steel!

Steel is such an important raw material because it is strong yet malleable. Steel can be heat-treated to make it harder or softer, flexible or stiff, ductile or brittle, depending on the application for which it is needed. In this project, we will consider the field of metallurgy and in the process, heat-treat steel to make a powerful and accurate rubber band shooting gun.

To do so, we will first anneal a piece of steel music wire to make it soft and malleable, then cut and form it to shape, and finally temper it to give the tough, springy characteristics a high-performance rubber band gun requires.

As you may already know, steel comes in various formulations and types, but the main ingredients are always iron and carbon. While other metals are sometimes added to the steel, it's the carbon that makes steel so special.

Unlike most other metal alloys, many things about carbon steel—hardness, toughness, elasticity, and so on—can be changed by applying heat in particular ways. Heat-treating makes some types of steel rock hard and brittle, which is good for cutting tools but not so good for skyscraper girders. Other heat-treating recipes make steels that are resilient and flexible, which are great for making coil springs but are the wrong choice for, say, fence posts.

Elasticity is what makes the Tempered Steel Rubber Band Shooter work. The gun shoots rubber bands that fly forth basically because they are stretchy and snap back into their original form after being elongated. The bigger the band and the greater the stretch, the farther the rubber band will shoot.

In order to hold and release the stretched band, we make use of the resiliency of tempered steel. In this project, we match up one type of springiness (rubber) with another (tempered steel) so we can make a fun little gizmo that's interesting to make and a lot of fun to shoot.

MATERIALS

- Music wire, ⅛-inch diameter × 36 inches long
- (1) Quart soybean oil (Soybean oil has a higher smoke point than most other vegetable oils)
- Heat shrink tubing, 8 inches long
- Rubber bands, assorted sizes and widths

TOOLS

- Safety glasses
- Heavy gloves

- Metal vise
- Propane torch
- ¾-inch steel pipe, about 1 foot long (¾-inch steel pipe has an outside diameter of about 1 inch)
- Large needle-nose pliers
- Slip joint pliers
- Hacksaw
- Bowl of water
- Drying cloth
- Steel wool
- Medium cooking pot with lid
- Hot plate
- Candy thermometer
- Grease-fire-capable fire extinguisher

TEMPERING THE STEEL

1. Don your safety glasses and heavy gloves.
2. Place one end of the wire in the vise. At this stage, the music wire is very stiff and difficult to shape because it is so springy. Light the torch and heat the wire until it's cherry red. Cherry red is a medium red color. If it glows bright orange, you've heated the metal too much. Move the flame slowly down the wire until the entire wire has been heat-treated. The heating and subsequent air cooling of the wire is a process called annealing, which makes the stiff music wire soft and malleable.

   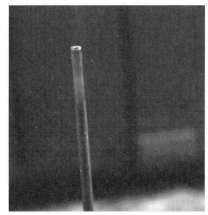

   5.1 Cherry Red Wire

3. Place the steel pipe in the vise and wrap the cooled wire around the pipe to make the top spring. Next, measure three additional inches and wrap the wire around the pipe again to make the bottom springs as shown in **diagram 5.2**.

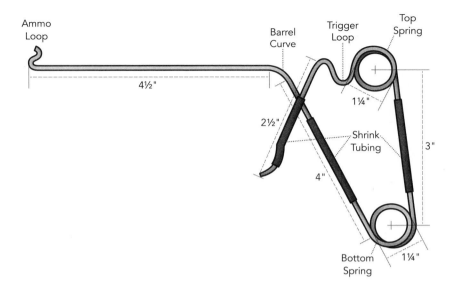

5.2 Shooter Wire Layout

4. Now you'll need to bend curves in portions of the wire to form the curves and indentations that hold the rubber band in place prior to launching. Use the pliers to push the now-bendable wire against the pipe in order to shape the arcs that form the ammo loop, the trigger loop, and the barrel curve shown in **diagram 5.2**. Cut off any excess wire with a hacksaw or rotary cutoff tool.

5. Heat the top and bottom springs with the torch until they glow bright red. Once hot, plunge them into a bowl of water. Dry the springs with a cloth, then clean off the scale with steel wool. At this point, the steel in the spring is hard and extremely brittle. Bending it even a little will cause it to break.

6. Do this step outdoors or keep a window open and ventilate the room well. Pour oil into the pot and heat to 400 degrees F, using the thermometer to check the temperature. Use caution here as the hot oil may smoke.

7. Reheat the round springs with the torch until they glow dull red. Stop heating and allow them to cool until the red color is completely gone, then plunge the springs into the oil. Turn off the hot plate and allow the oil and the springs to cool slowly to room temperature. This slow heat-treating process, called tempering, will make the steel springy and elastic.

5.3 Oil Quench

8. Wipe off the oil and position a piece of shrink tubing over the handle and trigger. The shrink tubing is flexible enough that you can easily thread it over the wire loop springs. Shrink the tube using heat from the hot plate or a match.

   If the heat-treating process worked properly, you now have a springy steel rubber band shooter. Load it by stretching a rubber band between the ammo loop and the trigger loop as shown in **diagram 5.2**. Pull the trigger and let the band fly!

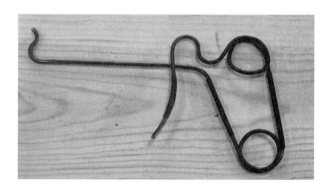

5.4 Tempered Steel Rubber Band Shooter

> ### *Safety Notes*
>
> 1. The smoke point is the lowest temperature at which flammable volatiles begin to rise from the surface of the soybean oil. The smoke point for soybean oil is quite high, around 450°F. Even though the oil doesn't ignite readily, you must take care to:
>   - not let the oil heat up past the smoke point
>   - have a lid handy to cover your pot quickly in the unlikely event the oil ignites
>   - keep a grease-fire-capable fire extinguisher close by
> 2. Use extreme care when handling the hot oil and wire.
> 3. Don't shoot rubber bands towards a person's eyes!

THE SCIENCE OF HEAT-TREATING

There are many types of heat treatment, but perhaps the best known are annealing and tempering. Annealing is a process in which steel is heated to a high temperature and then allowed to cool very slowly. Annealing steel generally increases ductility, that is, it improves the steel's ability to be formed into a shape.

Tempering is much like annealing except that the steel is first heated to a high temperature and then quickly cooled to make it hard and brittle, and then reheated to not quite so high a temperature and then slowly cooled. This makes the steel snap back into position after bending. We tempered the steel in our rubber band gun to make it tough yet elastic.

At the start of the Industrial Revolution, a somewhat brutal-looking piece of gear called the puddling furnace was the state of the art in the field of iron metallurgy. Iron workers loaded crude iron ingots into the furnace and then continually stirred the molten metal through a small hole. It may sound simple, but this took great skill and practice, and an experienced puddler was accepted in the industrial community as a highly skilled craftsman.

Puddler at Work

One historian stated that "only men of remarkable strength and endurance could stand up to the heat of the puddling furnace. . . . The puddlers were the aristocracy of the proletariat, proud, clannish, set apart by sweat and blood. Few of them lived past forty."

And although the process was hard, cumbersome, and expensive, the end product was still just wrought iron, which did not have the strength or utility of steel.

Steel is a mixture of iron and carbon (and sometimes other metals as well). There are many different types of steel, and the amount of carbon ranges from 0.2 to about 1.5 percent. Carbon, in just the right amount, makes steel harder than wrought iron but more ductile and forgiving than brittle cast iron. Steel is pretty great stuff, and that's why so much is made from it. From cars to I-beams to cannon barrels, steel is better than plain old iron in just about every way. That small inclusion of carbon makes a world of difference.

Early steel manufacturing was highly labor intensive, and the amount of fuel required to make even small amounts was mind-boggling. In the early 1800s, the typical procedure for making steel was to pack wrought iron bars in charcoal dust, one bar on top of another, and then place them into a furnace. The steel-maker would light a fire and let the heat build for many days. Slowly, the atoms of carbon in the charcoal would mix into the microstructure of the wrought iron. But the carbonization was hard to control and tended to occur haphazardly. So after the first heating cycle, the iron ingots would be broken into pieces and reheated for another lengthy period of time.

The result of this long and difficult process was a product called blister steel or crucible steel, either of which was an improvement over wrought iron but still fell far short of what was often needed by manufacturers. So the great hunks of blister and crucible would be heated yet again and then pounded with giant hammers until the carbon was evenly distributed and the steel was finally of usable quality. Because making steel in those times was a difficult and extremely expensive proposition, only a few parts would be made from such a dear material.

By the middle of the 19th century the railroads were booming; because of that, so was the market for iron and steel. The first railroads ran on wrought iron rails that were too soft to be durable. On some busy stretches, and on the outer edges of curves, the wrought iron rails had to be replaced every six to eight weeks. Steel rails would be far more durable, but there were no practical manufacturing methods to produce them affordably.

Then a smart fellow named Henry Bessemer came on the scene. Bessemer, an engineer and metallurgist, would over the course of his career receive 129 patents in a variety of engineering disciplines. But the invention for which he was knighted and the one that made him a very rich man was the one that involved turning iron and carbon into steel.

While looking for a way to increase the strength of iron cannon barrels, he discovered that carbon, if dissolved within molten pig iron, unites readily with oxygen. Knowing that, he determined that if he could blast a jet of air through molten iron, then he could convert the pig iron into much stronger alloy steel by using the air jet to precisely control the carbon content.

Sir Henry Bessemer

In 1856, based on the knowledge he gained in his laboratory experiments, Bessemer designed a machine that he called a converter. This was a large, napiform receptacle with holes at the bottom where large pumps could inject compressed air into the molten pig iron, emptying it of excess carbon and silicon in just a few minutes. Bessemer's process allowed him to make steel 200 times faster and cheaper than the then current state of the art in steel making.

THE SHAPE GUN

This chapter shows you how to build a powerful, high altitude potato cannon, which I call the Scientific High Altitude Physics Experimentation or SHAPE Gun. I think it's worth building for a variety of reasons.

First of all, it shoots higher and farther than any other cannon I've ever made. It's a bit of a challenge to build, but when you're done, you'll find it worth the trouble. And with it we can conduct some interesting science experiments. Besides its size, it is interesting because it has a unique ignition system that utilizes stun gun electronics to produce a powerful yet easy-to-control spark.

Let me be clear up front: building the SHAPE Gun entails some risk. This is a big cannon and it packs a lot of power, as in explosive power. So I've designed this gun with some additional safety measures. But nothing will keep you as safe as using it with common sense, and that means:

- **Don't look down a loaded barrel.**
- **Don't aim the gun at things you shouldn't.**
- **Don't use too much or the wrong type of propellant.**
- **Don't be careless with the stun gun ignition system**

While I have never had an issue making or using this big artillery piece, I provide no guarantees as to safety or efficacy. If you build this big guy, you proceed at your own risk.

MATERIALS

- 2-inch diameter schedule 40, non-PVC pipe, 10 feet long
- 4-inch diameter PVC pipe, 18 inches long
- (1) Stun gun*
- Wood block, 1 × 6 × 6 inches
- (2) ¼-bolts, 1½ inch long, 6 nuts
- 8 feet of two-conductor insulated 16-gauge wire
- Wood block, 1 × 6 × 8 inches
- (4) 1¼-inch deck screws
- (2) ¼-inch bolts, 2½ inches long, 6 nuts, 2 washers
- PVC primer and cement
- (1) 4-inch to 2-inch bell-shaped reducing connector
- (1) 4-inch female threaded adapter
- (1) 4-inch male pipe threaded plug
- Strapping tape, 1 roll
- Electrical tape
- Several large potatoes
- Long stick such as broomsticks taped together or a telescoping painter's brush holder
- Rope, 50 feet long
- (3) ground stakes
- Propellant†
- (2) Bricks and a piece of scrap wood
- Rag and nonvolatile cleaner

---

* In the states where they are legal, stun guns provide a huge spark for not much money. To find one to purchase, search online using the term "stun gun." If stun guns are illegal in your area or you choose not to use one for other reasons, see the notes associated with step 3, following.

† This is an aerosol spray with alcohol, butane, or other hydrocarbons in it. There are many products on the shelves of grocery stores that fit the bill; check the label and choose accordingly. Frequently, aerosol hair spray or dusting spray is used. Never use acetylene or Bangsite.

TOOLS

- File or rotary tool with sanding drum
- Electric drill and twist drill assortment
- ¼-UNC tap
- All-Purpose Measuring Tool Assortment
- Wire stripping and crimping tool
- Soldering gun
- Leather gloves
- Level and protractor
- Safety glasses

BUILD THE SHAPE GUN

The SHAPE Gun is constructed mostly from PVC pipe. You can find instructions for working with PVC pipe in the appendix on page 123.

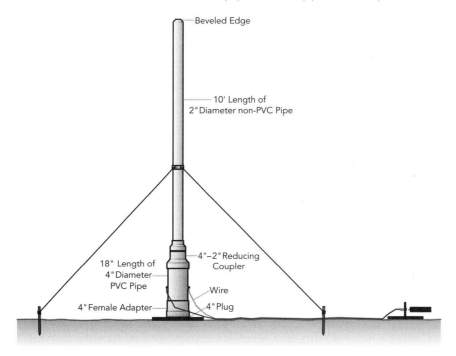

6.1 SHAPE Gun Assembly

1. **Diagram 6.1** shows the overall layout and assembly of the device. The first step is to fillet the edge of the barrel. Use a file or a rotary tool and sanding drum to taper one end of the 10-foot-long, 2-inch diameter pipe section so it forms a sharp

edge as shown. A clean, sharp edge is important because it has to cut the potatoes to shape as they are rammed into the muzzle.

2. Next drill electrode holes in the combustion chamber. Drill a 7/32-inch hole for the 1/4-bolt 4 inches from one end of the 4-inch diameter pipe (the combustion chamber). Drill a second hole of the same size also 4 inches from the same end, but 180 degrees away, as shown in **diagram 6.2**. Use a 1/4-UNC tap (Note, a 1/4-UNC tap is different from a 1/4-inch pipe tap) to thread the holes. If you're unsure as how to do this, review the explanation of tapping holes on page 26.

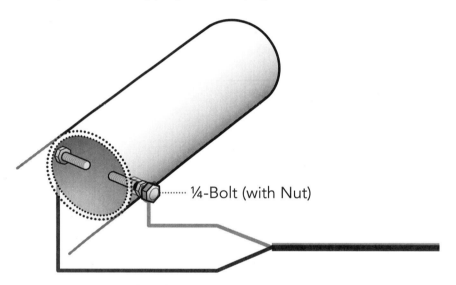

.... 1/4-Bolt (with Nut)

6.2 Combustion Chamber

3. * To build the stun gun ignition panel, turn off the stun gun, remove the battery, and then measure the distance between the protruding electrodes on your stun gun. Drill 5/16-inch holes that distance apart on the 1 × 6 × 6-inch block of wood. Insert the 1½-inch-long bolts into the holes. Strip 1 inch of the insulation from the 16-gauge wire. Wrap one end of the

---

* If stun guns are illegal in your area, or if you simply don't want to use one, don't worry! You can make an alternate (although not nearly as dramatic) ignitor using the piezoelectric igniter system we discussed in chapter 1 when building the Beverage Bottle Bazooka. Connect the two wires connected to the piezo igniter to the wires attached to the electrode bolts described in the following step.

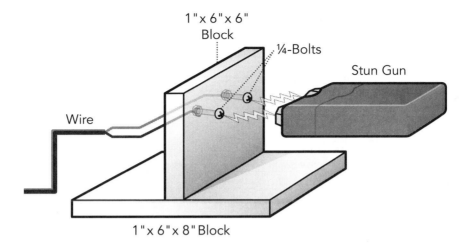

1"x 6"x 6"
Block

¼-Bolts

Stun Gun

Wire

1"x 6"x 8" Block

6.3 SHAPE Ignition Panel

stripped wire around each bolt, then fasten into positon with the nuts, as shown in **diagram 6.3**. Attach the firing panel to the 1 × 6 × 8-inch wood base using four 1¼-inch deck screws.

4. To install the electrodes, thread a nut onto each of the 2½-inch-long hex head bolts. Screw them into the holes you drilled in step 2. Put another nut onto each bolt, inside the pipe. See **diagram 6.2** again for reference. Position and adjust nuts as needed so there is a spark gap of approximately ⅓ of an inch between the bolt ends inside the barrel. This dimension can be adjusted to be larger or smaller depending on the power of your stun gun by positioning the nuts on the bolt.

Note: If you are using a piezo igniter instead of a stun gun igniter, be aware that the piezo generated spark is far smaller than the stun gun spark. So you will need to position the ends of the bolts much closer together in order to create a usable spark gap. Also, you may need to shorten the length of the 16-gauge, two-conductor wires in order to obtain a dependable spark.

5. It's time to solvent weld the parts together as shown in **diagram 6.1**, following the directions on the PVC cement bottle:

   a. Solvent weld the 4- to 2-inch reducing connector to the end of the 4-inch pipe closest to the electrode bolts. Clean the mating surfaces with PVC primer.

Then attach the reducing fitting carefully following the directions given on the PVC cement container.

b. Solvent weld the 4-inch threaded adapter to the other end of the 4-inch pipe. Screw the threaded plug into the threaded adapter opening.

c. Solvent weld the 10-foot-long, 2-inch diameter barrel to the corresponding opening in the 4- to 2-inch reducing connector.

d. Let the PVC solvent welded assembly dry in accordance with the directions on the PVC cement bottle.

6. Reinforce the combustion chamber by wrapping strapping tape around it. Strapping tape is pretty strong stuff, and while this doesn't eliminate danger, it does mitigate it somewhat. Anything you can do to increase your margin of safety is a good idea.

7. To make final electrical connections, remove an inch of insulation from the free ends of the wires. Wrap one of the exposed wires around each of the ¼-inch bolt electrodes. Solder the connections. Cover the soldered connection with electrical tape. Reinstall the stun battery into the battery holder.

SHOOTING FOR THE STARS

1. The SHAPE shoots a potato plug high (really, really high!) into the air. It is impossible to know exactly where the plug will fall. So you *must* perform this project only in a place where the potato will not damage any property or hurt people or animals when it falls. If you fire it vertically, be sure you have a covered spot in which to take shelter after firing so you don't get hit by the falling potato.

2. Carefully push a potato into the cannon from the muzzle end. The sharp cutting edge ground into the muzzle will cut the potato into a plug of the correct size. It's a good idea to wear leather gloves in case your hand slips. The potato must fit snugly on all sides of the muzzle. Any gaps will allow the expanding gas to "blow by" the potato. If that happens, the potato won't fly very far.

3. Use a long stick to push the potato plug 9 feet down into the barrel. You can tape a couple of broomsticks together or use a telescoping painter's brush holder for this.

6.4 Positioning the SHAPE Gun

4. Position the gun by making a tripod of rope and stakes as shown in **diagram 6.4**. Enlist a helper to hold the SHAPE Gun at the angle you wish to shoot. Pointing it straight up will shoot the potato to the highest altitude, while pointing it 45 degrees from the ground will shoot it the farthest distance. Once you decide on a direction and angle for the gun, pull the ropes tight and set the stakes in the ground. You can use a level and protractor to ensure the device is set at the angle you desire.

6.5 SHAPE Gun Base

5. Unscrew the 4-inch end cap. Direct a stream of aerosol spray (check the label to make sure it contains hydrocarbons such as alcohol, propane, or butane) into the firing chamber (the 4-inch diameter) cylinder where the electrodes are. I recommend a 2-second burst, but you will likely need to determine the best amount of propellant by trial and error. Using too much or the wrong sort of propellant is dangerous (e.g., never use acetylene or Bangsite).

6. Immediately replace the end cap and screw it on securely. Position the barrel at the desired elevation. Use bricks or wood to provide a secure base for the device.

7. Turn the stun gun to on. Observe all safety rules, including donning eye protection.

8. Press the stun gun electrodes against the ignition electrodes on the wooden board. Press the stun gun activation button.

   Steel yourself for the high energy liftoff that follows, and then enjoy the fruits of your labor! If all went according to plan, your SHAPE Gun just shot its projectile to an amazingly high altitude. But remember:

   - It is impossible to predict exactly where the potato will return to Earth.
   - Because the gun shoots to such great altitudes, it is often difficult to visually track the upward or downward path of the potato.

9. Aerosol propellants often contain chemicals that tend to gum up the inside of the cannon. Clean out the gun every few shots with a rag and nonvolatile cleaner. The end cap can get stuck from the gum. Use a wrench to unscrew it if necessary.

As Spider-Man's Uncle Ben once told him, with great knowledge comes great responsibility, so heed these safety notes:

- Stun guns are dangerous. Read and follow package directions to avoid unintended shocks.
- Use this device only in appropriate places.
- Never aim at anything you don't intend to shoot.
- Avoid being hit by falling projectiles.
- Unconsumed fuel may linger in the combustion chamber. Always treat the gun as if there is flammable vapor in the firing chamber.
- Potato cannons are dangerous. Use at your own risk.
- Wear safety glasses while constructing and operating the cannon.
- See "Staying in One Piece" on page xi for additional cautions.

HOW HIGH WILL IT SHOOT?

You can use simple Newtonian physics to determine how high the SHAPE will shoot, how far the potato will travel, and other interesting parameters such as the gun's muzzle velocity, which is how fast the projectile leaves the barrel.

The procedure for determining maximum altitude and the muzzle velocity is pretty simple if you've got a stopwatch and know which planet you're currently on.

The first step is to point the barrel straight up. Use a level to guarantee verticality and then use the ropes and stakes to stabilize the SHAPE in this position. Once that's complete, take out your timepiece. Now it's time to fire the cannon.

Please take note! You'll need to take cover immediately after firing to avoid being hit by a falling potato. Undertake this experiment only in a safe area where the potato will do no damage when it lands.

The science behind launching things vertically is actually pretty interesting. When you launch something straight up, it's a bit of a special case because the distance from the ground to apogee (the point when a projectile is farthest from the Earth) is equal to the distance from apogee to the ground. It is also a special case because the launch velocity is equal and opposite to the impact velocity. Together, this means that the time it takes for the projectile to reach its apogee from the ground is equal to the time it takes to fall from the highest point to the ground.

Knowing this, you can use your stopwatch or download an application to a smartphone to record the time elapsing between the firing of the cannon and the landing of the projectile. If you ignore the effects of air drag, then you can assume that half of the time the projectile will be going up, and half the time it will be falling.

Isaac Newton gave us a number of easy-to-use equations for figuring out how far, how fast, and how high something goes. Collectively, these equations are grouped under the heading "one dimensional particle kinematics." Don't let the rather daunting name scare you off from using these handy helpers. You can find a complete list of the equations in any beginning physics book. Believe it or not, I use them all the time for figuring out the answers to a number of ballistics and gunnery problems.

For example, the two simple equations here let you determine the capability of the SHAPE gun, specifically how fast the projectile exits the muzzle and how high it shoots.

Here are two key kinematics equations:

Height = Initial Velocity + ½ Acceleration × Time²

(In the case of our SHAPE, the Initial Velocity is 0.)

Muzzle Velocity = Final Velocity + Acceleration × Time

And in this case, the Final Velocity is 0. (That's because the projectile changes direction at the apogee, so at that instant the velocity of the potato is zero.)

I used these equations and conditions to create the following table. If you can determine the time it took for the projectile to go up and then come down, you can simply read the maximum height and speed from the appropriate column in the table.

The calculations behind the table below assume you are conducting this experiment on planet Earth, where the acceleration of objects due to gravity is 32.2 feet/second².

Total seconds, up and down, then divided by two	Maximum height in feet	Muzzle velocity in feet per sec	Muzzle velocity in miles per hour
1	16	32	22
2	64	64	44
3	145	97	66
4	258	129	88
5	403	161	109
6	580	193	131
7	789	225	153
8	1,030	258	175
9	1,304	290	197
10	1,610	322	219

For example, if you record a total time of 10 seconds between the firing of the gun and the landing of the projectile, then the time it took for the projectile to go up is 5 seconds. Therefore, from the table, the projectile reached a height of about 400 feet and left the cannon with a speed of nearly 110 mph.

HARP

Big, experimental guns like the SHAPE are super interesting to work with, so just imagine what it was like to work on the largest, farthest throwing gun in history.

In the 1960s, a group of Canadian scientists from McGill University in Montreal began work on HARP (High Altitude Research Project). HARP was a space capsule full of instruments designed to be shot straight up out of a cannon. The idea was to shoot HARP 90 miles up into the air in order to study what goes on in the high reaches of Earth's atmosphere. Shooting the HARP instruments out of a gun was hoped to be much less expensive than using a rocket to do the same job.

Obviously, the key element of such a project was building a gun with enough power to do the job. Such an ultra-powerful gun would be a monster indeed, bigger than the giant German Kaiser Wilhelm Gun (Paris Gun) of World War I and the even bigger top secret German V3 of World War II.

To engineer the gun, the HARP brain trust turned to a well-known gun designer named Gerald Bull. As a young man, Bull became fascinated with the technology of giant guns. Through intense study and hands-on experimentation, he ultimately became one of the world's foremost experts in the field of supersized cannons.

Bull convinced the US Navy to provide him with two barrels from Mark 7 battleship guns, which he connected end-to-end to create a single barrel of breathtaking length. With the barrel being so long and the instrument-filled projectile weighing only about 10 percent of what a normal 16-inch artillery shell did, the gun could shoot payloads literally into outer space. The huge cannon was shipped to Barbados in the Caribbean Sea. Its location, close to the equator, was just what the HARP scientists needed for their experiments. The big gun was surprisingly tough considering its odd construction. It shot a variety of large projectiles hundreds of times.

Later, a similar gun was constructed on US Army's Yuma Proving Ground in Arizona, and that was where the really high flying records were made. It was there that HARP engineers proved that things could be shot beyond the atmosphere of the Earth.

In the autumn of 1966, the world's largest artillery piece fired a 185-pound payload 111 miles high. (The altitude of 100 miles is called the Karman Line and is commonly considered the boundary between Earth's atmosphere and outer space.) That world altitude record still stands. The 1,225 pounds of powder that it took to do the job was perhaps the greatest amount of propellant ever loaded into a gun.

After the end of the HARP program, Gerald Bull went on to design 155 millimeter howitzers for the Canadian and South African militaries. These were excellent artillery pieces and are still used by military forces around the world. But unfortunately for him, that wasn't enough for Gerald Bull. He was the kind of man who was so driven to continue his invention and research of large guns that he would accept a job from whomever would provide it. In the 1970s, Bull agreed to work for Iraqi dictator Saddam Hussein on Project Babylon, which was to be a supergun that utilized all the technological breakthroughs that Bull gleaned from Project HARP.

The Iraqi supergun was estimated to be capable of firing a 350 millimeter projectile more than 600 miles, far enough to hit targets all over the Middle East. On March 22, 1990, an assassin shot Bull a half dozen times in the head outside his apartment in Brussels, Belgium. The identity of Bull's killer remains unknown.

CAST METAL CANNONRY

Sometimes, a model can be a good as the real thing, or even better. For example, consider the popularity of wargaming and the metal models that the pastime is based upon. Made of relatively soft metal cast into imaginative designs, wargaming figures are the stars of many table and board games.

What better way to ready a cannon than to cast the actual thing yourself (well, at least a scale model of it) using techniques and methods similar to those of the armorers of past armies?

Miniature wargaming is basically a strategy game that involves a number of miniature soldiers, typically sporting weapons and armor. Governed by a complex set of rules, it is often played on a surface modeled after a specific battlefield. The first wargamers were 19th-century German military strategists, whose purpose was to predict real-life battle scenarios by positioning armies using the laws of probability and chance through a game they called *kriegspiel*. Soon, other people realized that doing so wasn't just a way to wage real wars, but it could be turned into a game for fantasy military strategists as well.

You might be surprised to find out that it is not difficult to produce your own cannon models out of metal, at least for the simpler designs. Certainly, complex designs require quite a bit of casting and molding knowledge, but a model cannon can be made with easy-to-obtain materials. It takes a little bit of patience and practice, but reasonably good models are easily made. And I believe the models you cast yourself, of your own design, are far more satisfying to own and game with.

Casting a miniature replica of a cannon or culverin is a great introductory project to home foundry work. (A culverin was a large and heavy cannon used by French armies in the 1500s and 1600s.) The model is made out of wood and metal, just as real cannons are. You're not locked into making a culverin, however. Feel free to exercise your imagination and make whatever variations to the look of your model that suits you.

I designed this project so you won't need to search about for any hard-to-find materials or tools. Of course, casting metal means you'll need heat and quite a bit of it, and you must handle molten metal with care and make certain not to touch any hot parts. But the temperatures of the metals are no higher than you'll find in your home kitchen, so if you can handle hot oil safely, you can likely do this safely as well.

We'll begin by casting a cannon barrel from a wooden pattern. Then we'll cast some wheels and finish the project by building a wooden gun carriage to mount the cannon.

MATERIALS

- (2) ⅜-inch × ¹¹⁄₁₆-inch half round wood moldings, 3 to 4 inches long
- Shellac
- (1) Bag of commercial grade fine sand (Although you need only a few pounds, sand is typically sold in 50-pound bags. Luckily, sand is, well, dirt cheap.)
- (1) Pound of fine bentonite clay (You can find bentonite at health food or wine-making supply stores, or search online.)
- Water, about 3 ounces
- (1) Electrical switch box extension ring, 4 inches square and about 1½ to 2 inches deep. (You'll find these in the electrical

section of the hardware store or on the Internet by searching "Raco 8203.")

- (1) 1 × 8 pine board, 8 inches long
- ¾-inch diameter round wooden dowel rod, 8 inches long
- (1) Roll of lead-free, flux-free solder, 8 to 16 ounces depending on the size of your cannon pattern*
- 2-inch diameter copper cap (Available in the electrical section of large hardware stores or online; it's used for capping electrical conduit.)
- Metal-compatible glue
- (1) ⅜-inch square dowel, 3 feet long
- Wood glue
- Optional: Paint or varnish

TOOLS

- Carving tools (e.g., carving knives, sand paper, files, or a rotary tool)
- Scale
- Mallet
- Needle or awl and tweezers
- Heavy gloves
- Safety glasses
- Electric hot plate
- Large pliers

CASTING YOUR CULVERIN

1. Begin by carving the ⅜-inch by ¹¹⁄₁₆-inch half round wood moldings into a cannon shape. I think a 3- to 4-inch-long cannon is a good starting point, but you can make it larger or smaller if you wish. You can use a knife, sandpaper, or a rotary tool. Work carefully and make the cannon as realistic

---

* While lead casting is easy and inexpensive, it is safer to use alloys that do not contain lead. Lead-free solder can be purchased relatively inexpensively at most home stores and can be melted on a hot plate. Note: Besides solder, there are other lower-melting-temperature alloys that contain safer-than-lead metals such as bismuth and indium available online. Visit www.rotometals.com and search for low-temperature, no-lead casting alloys.

looking as possible. In foundry terms, the wooden carving is called the pattern, and the final casting will only be as good as the pattern you carve. I recommend keeping the pattern fairly simple to start. You can add detail as you become a more proficient metal caster.

When you have completed the pattern, paint or spray it lightly with shellac to seal it.

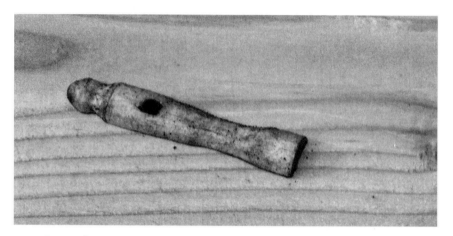

7.1 Carved Pattern

2. Prepare "green sand." Mixing the sand, bentonite, and water into the compound that foundry workers call green sand is the most critical part of the casting project. Although it's called green sand, it's often gray or black in color.

   To make green sand, the sand you start with must be fine and clean. Buy the finest (smallest grain size) sand you can find. If necessary, sieve the sand through a fine window screen to remove any grit or pebbles. When you're satisfied with the sand, thoroughly combine 40 ounces of sand and 5 ounces of bentonite clay. Moisten the mixture with about 3 ounces of water. The mix should be just wet enough to stick together when squeezed in your hand, but not so wet that water squeezes out.

3. Place the electrical extension box on the 1 × 8 board. Place the pattern flat side down on the board, inside the extension box. Fill the box with the green sand.

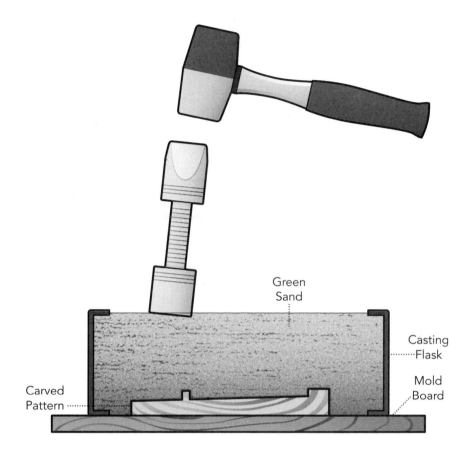

Green Sand

Casting Flask

Mold Board

Carved Pattern

7.2 Open Mold Casting

4. Using the mallet and ¾-inch round dowel, compact the sand around the pattern. Start at the corners and work toward the middle. Use medium pressure with the mallet to ram the sand.

5. When the sand is compacted, flip the electrical box over and carefully remove the pattern from the sand. You can use a needle or awl to wiggle it out. Inspect the mold. If you see any loose grains of sand inside the mold, blow them away or pick them up with tweezers. If the mold collapses or if the pattern does not release cleanly, repeat steps 3 through 5. If you consistently have trouble with the sand collapsing, try adding a little more water.

6. Don heavy gloves and eye protection. Place enough solder in the copper cap to fill the mold and then heat the cap and solder on the hot plate.

7.3 Melting Solder

When the solder liquefies completely, grab the pipe cap with the pliers and then pour the molten solder into the sand mold. Pour quickly and evenly into the open mold. The tricky part is pouring enough molten metal into the mold to fill it but not overfill it. Let the molten metal cool completely.

7. Remove the cooled casting from the sand mold and inspect. If you are satisfied with the quality of the piece, use sandpaper or a file to touch up the surface as required. (Some sanding and buffing will be needed to make the surface look smooth.) If you are not satisfied (and it will likely take you a few tries to get it right), cut up the casting into small pieces and remelt it in the pipe cap, and try again. That's the good thing about metal casting: you can melt your mistakes and try again!

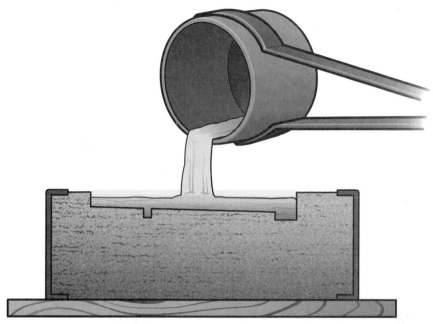

7.4 Pouring the Casting

8. Repeat steps 3 through 7, reusing the sand, to make the other half of the barrel casting.

9. Use sandpaper to sand the back sides of the cannon completely flat. Then use metal-compatible glue to fasten the two sides of the castings together. (A commercial foundry

7.5 Unfinished Castings

would cast the model as a single piece instead of two halves, but that adds some complexity to the process. But if you feel like experimenting, give it a try!)

10. Cut the ¾-inch diameter round dowel into four pieces, each about ¼ inch thick. Cast metal wheels using the ¾-inch round pieces as patterns, repeating steps 3 through 7 using the wheel pattern instead of the cannon patterns.

11. Assemble the model gun carriage from ⅜-inch square dowel pieces as shown in the photograph. The carriage is merely a platform to display the cast barrel, so you can experiment with various designs until you find one that suits you. Use glue to bond the wood pieces.

12. Attach the metal wheels to the carriage and position the cannon barrel on its carriage. Congratulations! Your model cannon is complete. You can finish the carriage with paint or varnish if you desire.

7.6 Finished Model Cannon

THE SCIENCE OF CASTING METAL

Metals (as well as many other materials such as water and butter) liquefy and solidify based on how much energy they contain. At higher temperatures, there's so much energy contained in the atoms and molecules of the substance that the chemical bonds holding a solid thing together break apart and it becomes liquid. When the heat is removed, the bonds reform and the substance resolidifies.

We use this property of matter to manufacture things. The process of pouring molten or liquid metal into a mold made of sand, metal, or ceramic is called metal casting. There are several different casting processes used in industry.

Die casting is a method where metal is injected under pressure into a strong, very precise mold known as a die. Die casting makes very accurate parts, but the dies and machinery required are expensive and complicated to make.

Another type of casting technique is called investment casting, and it is used to make intricately detailed products. Like die casting, it is complicated and expensive.

But the third method, called sand casting, is cheap and easy. It's the one we used to make our cannon model. We simply mixed up sand and bentonite clay and then rammed it around a wooden pattern inside a container. Once we removed the pattern, we poured in molten metal and waited for it to harden.

Of course, there is a big difference between our simple experiment and what happens in a commercial foundry. It takes a great deal of knowledge and experience to become proficient in sand casting to do it on a commercial level. But the basic ideas are easy to understand, and with a bit of practice, nearly anyone can cast simple items.

Any metal that you can melt can be used for casting. Of course, there are trade-offs involved, as some metals are much easier to work with than others. Some metals, like iron, are very strong but require very high temperatures in order to make them into a castable liquid state. Some, like lead, melt easily but are dangerous to work with because they are so poisonous.

For beginners, casting metals are usually made from different amounts of low-melting-point, low-toxicity metals like tin, bismuth, and indium.

H. G. Wells and Wargaming

In short order, a craze for waging war with toy soldiers developed among adults in Europe and especially in England. Some wargamers were well-known politicians and writers, including Robert Louis Stevenson, G. K. Chesterton, Winston Churchill, and most importantly, the science fiction writer H. G. Wells.

Herbert George Wells was the preeminent science fiction writer of the early 20th century. His best-known books include *The War of the Worlds*, *The Time Machine*, and *The Invisible Man.* One of his lesser known works is *Little Wars*, published in 1913 and considered by many to be the seminal work on wargaming. It was the first written rules for gaming using

THE WAR GAME IN THE OPEN AIR.

miniature infantry, cavalry, and artillery figures. Because of that, Wells is regarded by many gamers as the Father of Miniature Wargaming.

Today, the business of designing and producing figures for wargaming is a big industry. Hundreds of companies manufacture metal and plastic figures, and the span of time covered by the various wargames runs from ancient Greek to fantasy battles a thousand years in the future.

MUSHROOM CLOUD CANNON

The Mark 7 gun of World War II was one massive cannon. The biggest conventional gun carried aboard the biggest US warships, its 16-inch diameter barrel could fire a shell as heavy as 2,700 pounds. The huge ships known as Iowa Class battleships that used the Mark 7 carried nine of these big guys. All nine could be fired in a single broadside that totaled 12 tons of high explosive firepower. The Mark 7 didn't come online until late in World War II but was used extensively in the South Pacific, and it was also a player in Korea, Vietnam, and the Gulf War.

To propel the shell toward the target, the gun crew used gunpowder—and lots of it. Powering each shell were six silk-wrapped 100-pound bags of propellant that were rammed into the breach just before the shell was inserted. With the gun tipped to

Missouri **Mark 7**

Mark 7 Firing

a 45-degree angle, a 2,700-pound projectile could shoot 23 miles in about 80 seconds. The power in that powder was immense.

The Mushroom Cloud Cannon is a tightly controlled and much-scaled-down version of a powder explosion.

In this project, we'll carefully harness the energy contained in easily obtainable combustible powders and turn it into an exciting and extremely dramatic display of science at work. This project also goes by the name of the Coffee Creamer Fireball. That's because the energy that makes this project work comes from, believe it or not, dry powdered coffee creamer!

While the ingredients in coffee creamer differ by manufacturer, the main ingredients are frequently sodium caseinate and partially hydrogenated vegetable oil. Whether that's better for you than milk or cream I couldn't say, but I can tell you these coffee lighteners have more than enough easily ignitable, powdery fat in them to make an absolutely terrific pyrotechnic display.

A Mushroom Cloud Cannon is simply a device that creates an environment in which a finely powdered fuel is projected up into the oxygen-containing air. Once airborne, the fuel ignites, resulting in a large mushroom cloud as the fuel and air combine in a glorious display of high-energy chemistry.

The first coffee creamers hit the market in the 1960s, so at some time in the not-so-distant past, there was some pyrotechnic genius who first propelled some of this powder up into the air using a black powder lifting charge. The idea was born, and unfortunately, the name of the inventor is lost in the sands of time.

All in all, this is one of my favorite experiments. If you delve a bit under the cover, there's some interesting science to explore. This one is definitely not for the meek or the careless. However, with caution and proper preparation, it is an enjoyable and dramatic display of DIY chemistry. As always, follow the directions carefully, make sure a responsible adult is closely supervising the action, and before you begin, think over whether the risks and rewards associated with this somewhat edgy project make it a good one for you.

MATERIALS

- (1) 12-, 28-, or 46-ounce empty and clean steel (not aluminum) can. The bigger the can, the bigger your mushroom cloud will be. (Tip: Tomato juice often comes in 46-ounce steel cans.)
- Wooden board, 12 × 12 × 1 inches
- 1½ #10 bolt, washer, nut
- Black powder*
- Fuse†
- Tissue paper (The thin, stiff paper used to wrap gifts, not toilet tissue)
- Coffee creamer‡
- Bricks to bolster cannon

---

* Black powder comes in different granulation sizes. The size called FFg is about right for this project. Now, take heed here: black powder is different than products such as smokeless powder or Pyrodex. Do not use smokeless powder or Pyrodex. Use only black powder. Gun stores often sell black powder, especially if they cater to sportsmen known as muzzleloaders. If you can't find a store that sells black powder near you, search for "Goex black powder for sale" on the Internet.

† The type of fuse you need is called "visco cannon fuse." It is a red or green fuse made of a waterproof twisted fabric covering a core of black powder grains. It lights easily, stays lit, and burns at a 1 to 3 seconds per inch clip depending on the type you select. Buy fuse online. Good places to find it are www.Cannonfuse.com or www.Skylighter.com

‡ Interestingly, some coffee creamers work much better than others. A product that I can verify works well in this experiment is Cremora. It is sold mostly in the eastern part of the United States. In my experience, many other brands of creamer, such as Coffee-Mate, do not work as well. Some store brands (like Wal-Mart branded creamer) work pretty well, but you may need to experiment, as some store brands change their ingredients quite frequently. If all else fails, you can purchase Cremora online.

TOOLS

- Electric drill or awl
- Twist Drill Assortment
- Scissors
- Portable fire extinguisher
- Long-handled lighter
- Safety glasses

BUILD THE MUSHROOM CLOUD CANNON

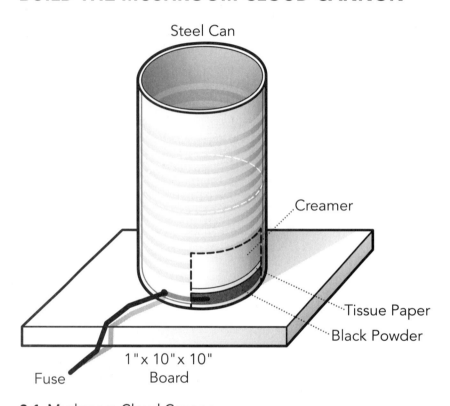

8.1 Mushroom Cloud Cannon

1. First, take a look at **diagram 8.1**. Using that as your guide, use an awl, a nail, or a drill and small drill bit to make a hole in the side of the can, ¼ inch above the bottom. Size the hole so it is just big enough in diameter to allow a length of fuse to be firmly inserted into the can.

2. Attach the can to the wooden board by drilling a ³⁄₁₆-inch hole through the center of each and connecting with a 1½-inch-long #10 bolt.

3. Cover the bottom of the can with a layer of black powder (BP) about ¼-inch deep. In my experience, a 3-inch diameter can

requires about ⅔ of an ounce of black powder spread evenly and lightly over the bottom. A 4-inch diameter can requires about 1¼ ounces of BP.

4. Insert a fuse into the hole in the can, making sure it is in contact with the black powder. Make sure the fuse length outside the can is long enough for you to get safely away before the device ignites. The fuse packaging should indicate its burn rate—that is, how many seconds per inch it burns. Seven or eight seconds is typically enough time to move to a safe place.

5. Use the scissors to cut a circle of tissue paper a bit larger than the diameter of the tube. Push it down the tube so it covers the black powder and fuse.

6. Fill the can a bit more than half full with coffee creamer. The cannon is ready to go!

KA-BOOM!

Find a safe place to try out your invention. The best place to use it is a location where you won't disturb neighbors or anybody else. The best results occur on calm, wind-free days or nights. Also, keep a portable fire extinguisher close by. Place bricks around the cannon so it can't tip over accidently.

Light the fuse with a long-handled lighter and immediately move to a safe place to watch the result. My guess is that you will like what you see!

8.2 Mushroom Cloud Cannon

8.3 Mushroom Cloud

Safety Notes

1. Cannon fuses burn intensely and hot. Use a long-handled lighter to keep your hand well away from the fuse when you light it.
2. The mushroom cloud will rise more than 20 feet into the air, and a thing like that gets noticed. Do this project only in an area where you will not disturb or alarm other people!
3. Be aware of what's downwind from the cannon, because that's where the cloud will float. While the cloud dissipates quite rapidly, it's important that you do not bother nonparticipants.
4. This project is for adults or to be supervised extremely closely by responsible adults.
5. And as you surely must understand by now, wear eye protection and try this at your own risk.

WHY DOES THE COFFEE CREAMER BURN SO INTENSELY?

This explosion packs a considerable punch. When flour, coffee creamer, or other types of ignitable dust float in the air, the rapidity with which it burns is based in part on the ratio of surface area to volume. The larger that ratio, the greater the likelihood of a rapidly burning or even explosive reaction occurring.

To understand why atomized coffee creamer powder makes a fireball when holding a match to a pile of the same stuff doesn't do anything, just consider the amount of fuel exposed to oxygen. When it's dispersed, fine particles of coffee creamer powder or flour or charcoal have a huge amount of surface area exposed to oxygen relative to their weight, so the stuff burns like mad. Lots of oxygen for each bit of fuel means fast combustion. And that's why dusty places can be so dangerous.

COMBUSTION CHEMISTRY

In all large cannons, from the largest World War II battleship guns to the Civil War Napoleon to the 24-pound British naval cannon we explored in previous chapters, gunpowder is rammed into the cannon before the projectile is loaded. When the powder ignites, the gun fires because the gunpowder itself contains a chemical, usually potassium nitrate, that supplies plenty of oxygen to the fuel to make it burn rapidly without any need for exposure to the air. That's what makes gunpowder so special: it provides its own internal oxygen for burning.

But many powders that don't contain an oxidizer will burn a bit like gunpowder if the conditions are just right, that is, somehow enough oxygen from the air can be provided to the fuel to promote intense burning. You may already be familiar with the "fire triangle," a graphical depiction of the three elements (fuel, oxygen, and ignition source) required for a fire to start. But there's another important, albeit less famous, polygon to know about. It's called the explosion pentagon.

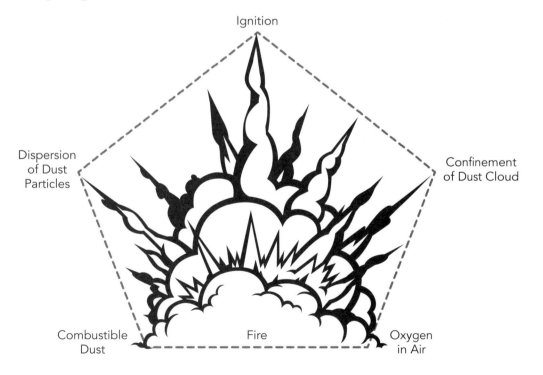

Explosion Pentagon

The explosion pentagon includes the three components from the fire triangle plus two more: a confined space and fine dispersion of the fuel. The dispersion of powder into a confined space will greatly speed up a normally slow-acting reaction. When these five elements come together, conditions are right for an explosive display of powder power.

Any combustible powder can cause a dust explosion. Dust explosions are an all-too-common industrial hazard. They can affect many different types of businesses, including grain elevators, metal finishing factories, and drug manufacturing plants.

Making a Mushroom Cloud Cannon gives you an idea of the chemistry of explosive powders. Each component in the cannon provides one of the five requirements of the dust explosion pentagon. The coffee creamer is the fuel, the fuse provides the ignition, the air in the atmosphere provides the oxygen, the upward motion caused by the rapidly burning black powder provides the dispersion, and the steel can provides the confinement.

SLINGSHOTS, HIGH AND LOW TECH

The slingshot is a far more recent invention than most people would imagine. Because it is so simple, it's hard to believe that it's only been around since the middle of the 19th century or so.

The reason for the slingshot's late arrival on the shooting scene is that it is, and always has been, powered by stretchy rubber bands. Until Charles Goodyear invented vulcanized rubber in 1844, such an item did not exist.

But with vulcanized rubber came rubber tires, and with them came the first rubber inner tubes. It did not take long for another now-forgotten genius to take a blown-out inner tube, cut it up, and attach it to a forked branch and start shooting beans.

It's easy to build a simple slingshot. A few rubber bands and a piece of hickory and you're pretty much set to go. But with a bit of thought and a few dollars in supplies, it's not difficult to go far beyond that rustic plaything and instead build a high-power, accurate, and dependable slingshot.

As you'll see, vulcanized rubber was, and still is, a mighty important chemical breakthrough, and it's the thing that makes the two projects described in this chapter, the Classic Slingshot and the Wrist Cannon, such high-performance shooters.

Both projects make efficient use of modern rubber's incredible toughness and energy storage capabilities. The first project is called the Classic Slingshot and, as its name implies, it is a simple and straightforward rendition of an old-timey bean shooter. But looks are a bit deceiving as there is a twist involved (both figuratively and literally) that makes this one perform better than the average slingshot.

The second project, the Wrist Cannon, is a bit more complex to build, but still, just about anybody can put it together in a few hours, start to finish. And the Wrist Cannon's performance is amazing. Powerful, accurate, and dependable, it's an absolutely top-notch slingshot.

But before you go any further with these projects, heed these warnings: Slingshots shoot with great power. Be smart and be safe by following this advice:

1. Wear safety glasses.
2. Slingshots are not toys. Children should always be closely supervised by adults if allowed to use a slingshot.
3. Both of the slingshots described here are powered by latex rubber tubing. Prior to and during use, inspect the tubing for any indication of wear or weakness. If you find any, replace the latex tubing immediately.
4. Do not shoot at hard surfaces or at the surface of a lake or pond. Watch out for ricochets.
5. Slingshot pellets can travel hundreds of yards. Know what's behind and close to your target. Place a backstop behind your target and make certain it has adequate strength to stop pellets. Never aim your slingshot at anything you don't want to hit.

The Updated Classic Slingshot

This is a new "twist" on a classic slingshot design. By twisting the half-round profiles of a pair of PVC forks, you'll get a device with a sturdy frame that resists bending and necking.

MATERIALS

- (1) 1-inch diameter PVC pipe, 11 inches long
- (1) Paracord (parachute cord or similar type of rope), 6 feet long
- (2) Pieces of ¼-inch ID latex rubber tubing, 9 inches long (You can experiment with different thicknesses and lengths of rubber tubing. Thicker walled tubing will be harder to pull back but will shoot farther.)
- Rubbing alcohol
- Dental floss
- Leather pouch, cut into a 1-inch × 2¾-inch shape
- Bean or pebble

TOOLS

- Handsaw or hacksaw
- Vise
- Heat gun
- Heavy-duty leather gloves
- (2) 2 × 4 blocks of wood, about 6 inches long
- Drill
- Drill bit set
- Rotary tool with sanding drum or file
- Utility knife
- Safety glasses

Optional: Methyl Ethyl Ketone solvent and rag (Read and follow all instructions on the Methyl Ethyl Ketone container.)

BUILD THE UPDATED CLASSIC SLINGSHOT

1. First, make the slingshot frame. If you'd like to, you can clean and remove the printing from your PVC pipe by rubbing it with a rag wetted with Methyl Ethyl Ketone. When the pipe is

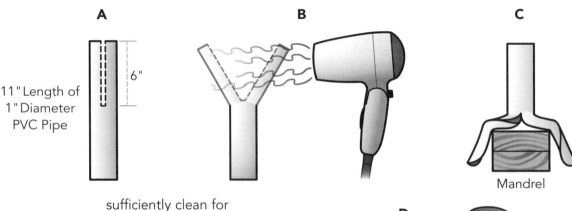

A

11"Length of
1"Diameter
PVC Pipe

6"

B

C

Mandrel

sufficiently clean for your purposes, let it dry. Cut the PVC pipe in half, longitudinally, for a length of 6 inches as shown in **diagram 9.1.A**.

2. Secure the uncut end of the pipe in a vise or other holder and heat the cut portion of the pipe with the heat gun until it softens, as in **9.1.B**.

3. Wearing heavy gloves (the plastic is still hot), separate the cut portions of the softened PVC and twist them into tight spirals. Then, while still hot and pliable,

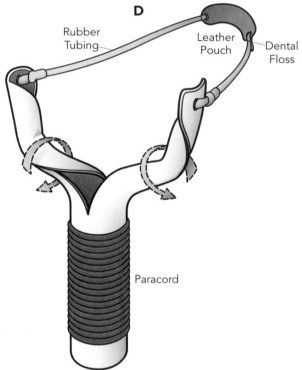

D

Rubber Tubing

Leather Pouch

Dental Floss

Paracord

9.1 Updated Classic Slingshot Assembly

shape the spirals into a U-shape using the two 2 × 4 blocks as a mandrel. Hold the two halves against the wood blocks until the plastic cools and the wings hold their new shape, as shown in diagram **9.1.C**.

4. Cut ⁵⁄₁₆-inch holes in the upper center of the wings as shown in **diagram 9.1.D**. Sand or file the edges of the holes.

5. Drill a ¼-inch hole in the base of the frame. Insert the paracord and tie a knot in the tube interior. Wrap the paracord around the frame tube tightly and bind by drilling another hole in the frame, pulling tight, and tying off.

6. Insert one piece of rubber tubing through each ⁵⁄₁₆-inch hole in the slingshot's wings.

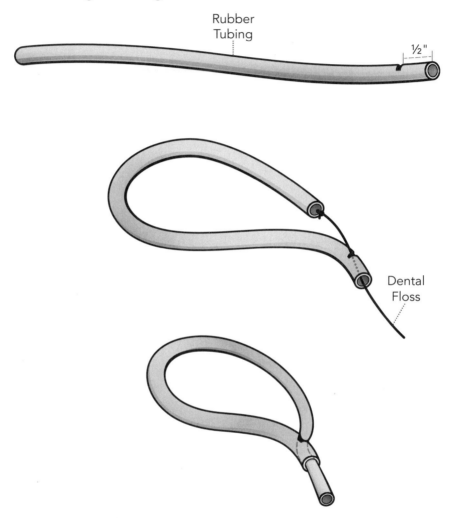

9.2 Power Band Loops

7. Cut a small lateral slit about ½ inch from one end of each latex rubber tube.

8. Lubricate the slits with rubbing alcohol. Tie a piece of dental floss around the opposite end of the tube, then insert the thread through the slit and out the open end of the tube. Pull

the dental floss and tube through the slit so the latex tube forms a loop. Pull until the loop is drawn tight around the wings.

9. Punch holes in the leather pouch as shown in **diagram 9.1**. Insert the nonlooped end of the rubber tubing through the holes. Fold over and tie securely with dental floss. Test the pouch connection by pulling back on the slingshot.

Your Updated Classic Slingshot is ready for use!

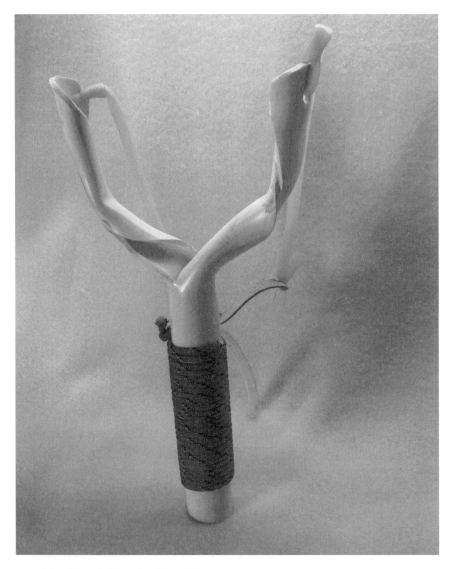

9.3 Updated Classic Slingshot

SHOOTING YOUR ARM-POWERED ARMS

To use your new slingshot, place a projectile such as a bean or pebble in the pouch. Don safety glasses; pull back on the pouch by stretching the rubber tubing. Take aim, and release. Remember, your slingshot is not a toy and shoots with considerable power.

Here's a trick question: Who founded the Goodyear Tire and Rubber Company?

If you answered Charles Goodyear, sorry, try again.

Although he did get rich making and selling tires, Charles Goodyear may have been the most dogged and unrelenting solo inventor of the 19th century's golden age of invention. But for a fellow with such a famous name, many things about him are misunderstood or little known.

First, Charles Goodyear did not start, work at, or even know of the giant industrial concern called the Goodyear Tire and Rubber Company. The company, which was named in Goodyear's honor, was founded in 1898 by industrialist Frank Seiberling about 40 years after Goodyear died.

Second, the incredibly valuable contribution to science and technology he made wasn't the invention of a thing like a telephone or a lightbulb but rather the invention of a process now known as rubber vulcanization.

And third, the way Goodyear came across it wasn't though his great scientific knowledge or precise experimental technique but rather a lucky accident.

The story begins 65 years before Goodyear's birth, when in 1735 the first great international scientific expedition, called the French Geodesic Mission to the Equator, traveled from Paris to Peru in order to accurately measure the shape of the Earth. While they were there, the scientists observed the native peoples tapping tropical trees to obtain latex.

The explorers brought some of the stuff back to Europe, where it was regarded as interesting but of little practical use. The one thing it could do rather well was to remove pencil marks when rubbed briskly on paper.

Seventy-five years later, though, latex rubber became all the rage. In the winter of 1820, a rubber fad swept across America when millions of people bought rubber-coated boots to keep their feet dry. But the fad ended as abruptly as it started when consumers found that a single summer of hot weather turned their rubber shoes to mush. Nearly all of the rubber manufacturing companies closed. Natural rubber clothing just wasn't durable or practical.

At that point Charles Goodyear entered the scene. Goodyear thought that if he could figure out a way to toughen the rubber chemically, he would have a product that people would buy. Although Goodyear knew almost nothing about chemistry, engineering, or business, he was, like the substance he was trying to make, resilient and tough.

He began to experiment with latex. He mixed in witch hazel, magnesia, even cream cheese in attempts to turn sticky, soft latex into durable, tough rubber. Nothing worked. He got close a number of times, but he was not able to develop a repeatable process that would turn rubber into a useful raw material.

He kept trying until he had spent all his money, so he borrowed some. He spent that and borrowed more. Eventually, he and his family were broke and living on the charity of his friends. Things looked bleak.

Then an accident changed things. On a cold winter day in 1839, Goodyear accidently brought a piece of rubber that he had treated with sulfur into contact with a hot stove. The inventor looked at the piece in amazement. Something incredible had happened. Goodyear wrote:

> I carried on some experiments to ascertain the effect of heat on the [sulfur-treated latex rubber]. I was surprised to find that a specimen, being carelessly brought into contact with a hot stove, charred like leather. Nobody but myself thought the charring worthy of notice. However, I directly inferred that if the charring process could be stopped at the right point, it might divest the compound of its stickiness throughout, which would make it better than the native gum.
> Upon further trials with high temperatures I was convinced that my inference was sound. When I plunged India rubber into melted sulfur at great heats, it was perfectly cured.

This rubber fragment did everything that natural latex could not. It had become, in modern parlance, vulcanized. It was tough and durable in hot weather and stayed flexible in cold weather. This, Goodyear knew, was a product with a huge future.

In the winter of 1841, things were looking up for Goodyear. His new process was an astounding success, and money started to come his way. During the years of experimentation, he had racked up $35,000 in debts (about $750,000 in modern dollars). He was now able to pay off all he owed within a few years.

But Goodyear was not a capable businessman. While he did obtain a patent for the vulcanization process in 1844, he licensed the process at rates that were far too low for him to make money. Worse, when patent infringers stole his work, he spent more on his attorney's fees than he was able to recover from the pirates.

He spent the rest of his life attempting to make good his dream of becoming a millionaire rubber manufacturer. Goodyear staged magnificent displays showcasing rubber products, including furniture, floor coverings, and jewelry, at London and Paris exhibitions in the 1850s. But while in France, his French patent was cancelled and his royalties stopped, leaving him with outstanding bills he could not pay. Goodyear was thrown into a debtors' prison.

Charles Goodyear

When he died in 1860, Charles Goodyear was $200,000 in debt. Posthumously, royalties on his process started to roll in. His son Charles Jr. later made a fortune manufacturing shoemaking machinery. It's a shame that Charles Sr. never enjoyed the financial success his invention ultimately provided for his family.

The Wrist Cannon

Not only simple to build and easy to use, the Wrist Cannon is also accurate and dependable. It incorporates a unique design featuring a rope wrist brace, making it much easier to hold and aim when the power band is stretched than traditional, nonbraced slingshots.

In addition, it uses a single, uncut power band to attach to the ammunition pouch instead of two separate pieces of tubing. This makes it easier to equalize tension between the wings of the slingshot, making for a truer, straighter projectile trajectory.

MATERIALS

- (4) ½-inch diameter 90-degree PVC elbows
- (4) ½-inch diameter PVC tee fittings
- (6) ½-inch diameter PVC pipes, each 1½ inches long
- (4) ½-inch diameter PVC pipes, each 4¼ inches long
- (2) ½-inch diameter PVC end caps
- Vinyl tape
- 34 inches of ¼-inch ID rubber latex tubing
- Leather pouch, cut into a 1-inch × 2¾-inch shape
- Dental floss
- 8 feet of paracord or other rope
- Bean or pebble

TOOLS

- Needle-nose pliers
- Scissors or knife
- Safety glasses

Optional: Methyl Ethyl Ketone solvent and rag (Read and follow all label instructions on the Methyl Ethyl Ketone container.)

BUILD THE WRIST CANNON

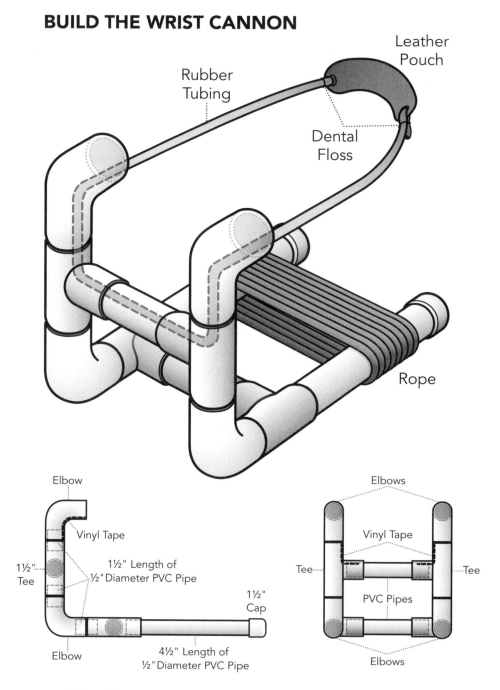

9.4 Wrist Cannon Assembly

1. Refer to **diagram 9.4**. Begin by assembling the left and right
   wings. You can rub the PVC parts with a rag soaked with Methyl
   Ethyl Ketone to remove the printing if you wish. For each wing,

attach the top elbow that holds the tubing to a tee fitting. Then attach that to the second elbow and then to the second tee. Use the 1½-long pieces of pipe as the connectors. You do not need to use PVC cement and primer on the joints of this device. Finally, attach a 4¼-inch-long pipe to the remaining open hole on the bottom tee and place an end cap on the open end of the pipe.

Place a piece of vinyl tape inside the elbow and upper tee fittings to provide a smooth, nonabrading surface to support the rubber band.

2. Insert the rubber tubing through the opening of the elbow fitting on the left wing and thread it so it comes out of the middle socket of the tee fitting below it. Pull the tubing through a 4¼-inch pipe and then press the pipe into the middle socket of the tee and push the connection tight.

3. Pull the tubing though the middle socket of the tee fitting on the right wing and then pull it up and out of the elbow tee fitting above it. You may need to use needle-nose pliers in order to reach down through the elbow tee's socket to grab the tubing. Push the connection between the pipe and tee tight.

4. Insert the remaining 4¼-inch pipe into the sockets of the remaining two tees and push tight.

5. Punch holes in the leather pouch as shown in **diagram 9.4**. Insert the ends of the rubber tubing through the holes, fold over, and tie securely with dental floss. Test the pouch connection by pulling back on the slingshot.

6. Using rope or paracord, tie a clove hitch around the bottom right 4½-inch pipe and then wrap the cord loosely between the two 4½-inch pipes as shown in the diagram. Finish wrapping the paracord by tying a clove hitch around the 4½-inch pipe.

SHOOTING YOUR WRIST CANNON

Put on safety glasses. Grasp the bottom crosspiece in your hand with the rope brace above your hand. Place a projectile such as a bean or pebble in the pouch. Extend your arm. Pull back on the pouch by stretching the power band. Take aim and release the

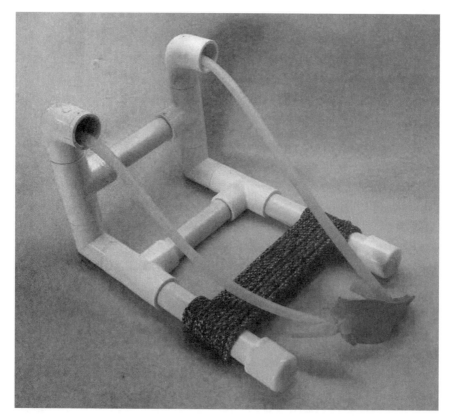

9.5 The Wrist Cannon

pouch, taking care to do so as smoothly as possible. With practice, you and your Wrist Cannon are capable of shooting with great accuracy. Check the power band for wear frequently.

THE SCIENCE OF SLINGSHOTS

To understand how and why the slingshots described in this chapter shoot with so much power, you first need to understand that there are two types of energy: potential and kinetic.

Kinetic energy is energy that a moving object possesses. A major league fastball, an arrow flying through the air, or a spoon falling from a kitchen table are all examples of objects that possess kinetic energy simply because they have mass and they are moving.

The other type of energy is called potential energy, which is energy stored in a thing because of its position. A drawn bowstring and arrow aren't moving, but they have a lot of potential energy because of the forces in the bow. A spring stretched out and held immobile isn't moving, but there's potential energy in the spring because of the spread out positon of its coils.

In a slingshot, when you pull pack on the rubber tubing, you input potential energy into the elastic rubber tubing, and when you let it go, kinetic energy is released in the form of a fast-moving projectile.

The nature of elastic systems like this one was first explained in the 17th century by an English scientist named Robert Hooke. Hooke determined that if you knew a little about the mechanical properties of a spring (or really any elastic thing), you could predict how much force it would take to pull it back a given distance and how much energy would be freed when the spring was released.

While that might sound a bit ho-hum, the brilliant Mr. Hooke went much further. Many elastic systems like metal springs, a drawn bowstring, or a rubber-band-powered slingshot can all be modeled by a mathematical formula. This formula, now called Hooke's Law, says that the amount of force released by a spring is linearly proportional to the amount it is stretched.

Written out in equation form, Hooke's Law looks like this:

Force = Constant × Distance, where

Force is the amount of force it takes to pull back the spring;

The constant is some unchanging number corresponding to the material the spring is made from;

Distance is the length that the spring or bowstring or tube is pulled back.

This equation shows that the amount of force it takes to pull back on the pouch is proportional to the distance pulled back times some constant. So if you have a spring with a constant of 1, it would take 2 pounds of force to pull a spring back 2 inches, 4 pounds of force to pull a spring back 4 inches, and so on.

But the really interesting part of Hooke's Law is that when you pull back on the pouch, the amount of energy stored in the rubbery slingshot sling goes up not proportionally but exponentially. Put another way, the amount of potential energy input and the amount of kinetic energy released is equal to the square of the distance the spring was pulled back times a constant that depends on the properties of the material from which the spring was made.

The potential energy formula looks like this:

Potential Energy = 0.5 × Constant × Distance × Distance

Well and good, you might be saying to yourself, but so what? Actually, this is pretty important. From this formula you can see that the amount of potential energy imparted to a bean or pebble in the sling of your slingshot rises dramatically as you pull back on the tubing. If you double the distance you pull back on the tubing, you impart not twice but four times the energy to your projectile!

These formulae explain why slingshots are so powerful and so useful. A well-made slingshot with good quality bands can shoot with amazing force and accuracy.

THE NEAR-SUPERSONIC PING-PONG BALL LAUNCHER

This chapter shows you how to build a projectile launcher that shoots with mind-boggling muzzle velocity. Maybe not high velocity rifle bullet speeds, but one with a muzzle velocity that's very likely higher than any other cannon you have ever built. By combining air pressure, mass production technology, and good design, you can make a device that delivers an unhittable table tennis shot!

This device shoots a Ping-Pong ball at high speed. It's not hard to build this gun, but I don't recommend you actually try to incorporate it in your table tennis practice unless you dial way back on the air pressure. Believe me, with more than 8 psi in the pressure reservoir, you don't want to be anywhere near the muzzle end of this shooter when it fires. Consider this a science experiment and not an addition to your arsenal of table tennis techniques.

If you crank the pressure inside the reservoir to 35 psi, the ball rockets out faster than Moody's goose. In fact, it's so fast that even something as lightweight as a Ping-Pong ball, weighing in at a paltry 2.7 grams, possesses an amazing (and dangerous) amount of kinetic energy.

Use your head, wear eye protection, and do not stand in front of the gun (whether loaded or unloaded) when the gun is pressurized. The pipe manufacturers print the highest allowable pressure on the side of the pipe, and you should never use more than a third of that as the highest allowable pressure. And remember, proceed at your own risk.

While we're on the subject, you should note that using PVC as a container for pressurized air is not sanctioned by the companies that make PVC pipe. While I have never had a problem with it at the relatively low pressures described here, PVC pipe manufacturers maintain that this sort of activity can be dangerous. So use PVC at your own risk or substitute steel pipe for the plastic.

This project, like the Mack-Mack Gun described in chapter 3, requires you to drill and tap holes in PVC pipe. You might want to review the drilling and tapping instructions given on page 26 or seek guidance from an experienced friend if you've never tapped a hole. But the good news is that because you're tapping holes in plastic, it's a relatively easy job.

MATERIALS

- (1) 1-inch inline sprinkler valve
- (1) ¼-inch diameter NPT brass hex or regular nipple, 1 inch long
- (1) Compressed air palm-type blow gun (Make sure the one you buy has a ¼-inch NPT air connection.)
- Epoxy or hot glue
- (1) 3-inch socket-style PVC end cap
- (1) 3-inch diameter PVC pipe, 18-inches long
- (1) ¼-inch NPT air tank valve*

---

* Tank valves are available at hardware and home stores that have reasonably large inventories. You can also find them online by searching for "air tank valve" or order from an industrial supply company like McMaster-Carr or Grainger. The appendix on page 121 provides information on how to contact industrial supply companies.

- (1) ¼-inch NPT, 0 to 60 psi pressure gauge
- (1) 3- to 1½-inch reducer
- (2) 1½-inch spigot × 1-inch female pipe threaded reducing bushing
- Strapping tape
- (1) 1½-inch diameter PVC pipe, 5 feet long
- (1) 1½-inch socket-to-socket coupling
- (2) 1-inch NPT close iron pipe nipples
- Ping-Pong balls
- Long stick, such as a broomstick

TOOLS

- Electric drill
- Twist Drill Assortment
- ¼-inch NPT pipe tap and handle (You want a ¼-inch pipe thread tap, not a UNC tap.)
- Adjustable wrench
- Pipe thread sealant
- PVC cement and primer
- Safety glasses

MODIFY THE SPRINKLER VALVE

Water sprinkler valves are used in lawn irrigation systems. We're going to modify one so it acts like a relay. A relay is a device that uses a small amount of fluid pressure to control a much larger amount of fluid pressure. We will adapt the sprinkler valve so we can control the big volume of air in the air reservoir by releasing a small amount of air with a hand-operated lever.

1. Disassemble the sprinkler valve. On some models the top unscrews, and on others you'll need to remove screws or bolts holding the top to the body.
2. Drill a ⁷⁄₁₆-inch hole in the center of the top, which is the correctly sized hole for tapping the ¼-inch NPT pipe thread on the air gun. Insert a ¼-inch NPT tap into the tap handle and cut screw threads into the hole you just made. If you need a reminder on how to tap a hole in plastic, review the instructions on page 26.

10.1 Modified Sprinkler Valve

3. Apply pipe thread sealant to one end of the ¼-inch diameter, 1-inch long NPT brass nipple and screw it into the tapped hole you just made.
4. Apply pipe thread sealant to the other end of the ¼-inch diameter NPT brass hex nipple and screw on the ¼-NPT opening on the blow gun.
5. Block the solenoid air hole (see **diagram 10.1**) by covering it with epoxy or hot glue. Reattach the top to the valve body.

CREATE THE PRESSURE RESERVOIR

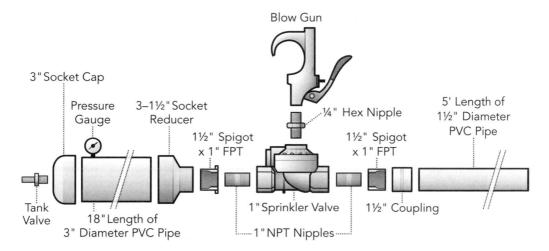

10.2 Near-Supersonic Ping-Pong Ball Launcher Assembly

1. First, refer to **diagram 10.2**. Drill a 7/16-inch hole in the center of the 3-inch PVC end cap and in the 3-inch diameter pipe. Use the ¼-NPT pipe tap to cut ¼-NPT threads in the hole. Apply sealant and then insert the ¼-inch tank valve in the end cap, and the pressure gauge in the pipe.

2. Assemble the reservoir as shown in **diagram 10.2** using PVC cement. Follow all cement and primer label directions carefully. Construct the reservoir assembly by cementing together the parts shown on the left-hand section of **diagram 10.2**. This consists of the 3-inch cap, the 3-inch diameter pipe, the 3-inch to 1½-inch reducer, and the 1½-inch by 1-inch reducing bushing.

3. Neatly wrap the pressure reservoir with strapping tape.

BUILD THE BARREL

1. Now build the barrel assembly, shown on the right-hand side of **diagram 10.2**. Use PVC primer and cement to connect the 1½-inch diameter barrel, the 1½-inch coupling, and the other 1½-inch by 1-inch reducing bushing.

FINAL ASSEMBLY

10.3 Near-Supersonic Ping-Pong Ball Launcher

1. Screw the modified valve assembly into the 1-inch female pipe thread opening on the pressure reservoir. Air leaks are your enemy! Use pipe thread sealant to prevent leaks.
2. Screw the valve assembly into the 1-inch female pipe thread opening on the barrel assembly, using sealant to prevent leaks.
3. Use the compressor to pressurize the air in the reservoir to 15 psi. Watch the pressure valve to see if any air leaks. If it does, you'll need to find the leaks by spraying soapy water on all your connections and looking for bubbles. Then press the trigger on the blow gun and test the operation of the valve. Release the air when done.

HOW TO MAKE A 300 MPH TABLE TENNIS SERVE

The Near-Supersonic Ping-Pong Ball Launcher is a wonderful example of Boyle's Law. A lot of pressure in a small volume (the reservoir) is let loose in a larger area (the barrel). The volume of gas expands nearly instantaneously, and that gas pushes hard on the Ping-Pong ball. Since the ball is constrained inside the barrel, there's a lot of force pushing on it and that increases the ball's velocity as it hurtles down the barrel. The bottom line is that you'll see some impressive muzzle velocity readings if you measure the ball with a ballistic chronograph or take high-speed movies. I've recorded velocities exceeding 300 miles per hour. That's nearly .5 Mach!

Let's see what this thing will do, but before we do, review the following commonsense safety rules.

Safety Notes

- Wear safety glasses.
- This cannon shoots with enough force to cause great harm. Use at your own risk.
- Do not look down the barrel.
- Always know where the barrel is pointing. Aim only at targets you intend to hit.

1. Insert the Ping-Pong ball and use a straight stick (like a broomstick) to push it as far down the barrel as it will go.
2. With the air compressor, pressurize the air reservoir to 30 psi. You must limit the pressure used. The higher the pressure, the more dangerous the device becomes.
3. Once you are sure your gun is aimed at a suitable target, such as a wall ten feet or more from the muzzle, press the trigger on the blow gun. The gun will fire with a loud report and a whopping amount of power!

ROBERT BOYLE

Many history of science scholars make the case that Robert Boyle was the most important scientist ever born in Ireland. In fact, a few scholars have argued that of all the people ever born in Ireland, Robert Boyle had the greatest impact on the course of human history. (Take that, Bono.)

Robert Boyle was born in County Waterford in 1627 and is often regarded as the father of modern scientific inquiry. He made his name not from making bold theoretical suppositions but through meticulous testing and experimentation. He was very, very good at this, and his best-known achievement is the principle that bears his name, Boyle's Law, the first step forward in the field of thermodynamics.

How important was Boyle in the annals of scientific history? Well ponder this. Just two days after Boyle's funeral on January 7, 1692, the most influential men in England, Samuel Pepys and his fellow diarist, John Evelyn, sat down at Pepys's house to decide whom they would back as the new leader of English science.

Can you guess whom they chose to assume Boyle's mantle? None other than Isaac Newton. This shows how the late 17th-century savants were unanimous in their choice of Boyle and Newton as the two great leaders of Enlightenment-era science.

Part of the reason Boyle chose to devote his life to science rather than to making money was that he was the son of Richard Boyle, the first Earl of Cork and one of the most powerful landholders in 17th-century Ireland. Lord Richard is said to have been a ruthless and sharp-dealing robber baron, a wildly successful Elizabethan land thief of gargantuan ambition. Seeing his father's ways, son Robert became

Robert Boyle

wary of the motives of commerce and had no interest in following in those footsteps.

Despite his Irish roots, he became frustrated at his inability to make progress as a scholar of chemistry in Ireland. So in 1655 Boyle left for Oxford in England to pursue his work. Boyle had his eureka moment in 1662 while he was experimenting in his home. He realized that nearly all gases behaved a certain way when he put them inside closed containers and then changed the size of those containers. And he found that the way they acted was mathematically describable.

If you put a cap on a bottle and push it down, the pressure inside the bottle increases when the size of the bottle decreases. This idea, that when one variable in an equation goes up, the other goes down, is a mathematical relationship called an inverse relationship.

As volume increases,
pressure decreases

As volume decreases,
pressure increases

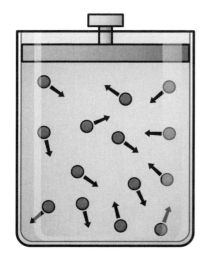
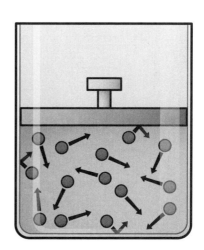

$$P_1V_1 = P_2V_2$$

Boyle's Law

It turns out that the relationship between the pressure of a given amount of gas molecules and the volume they occupy is an inverse relationship.

Boyle's Law is written mathematically as $P_1V_1 = P_2V_2$

P stands for pressure and V stands for volume.

There are thousands of examples of Boyle's Law in the natural world. Perhaps you're aware that fish that live in very deep seas die when they are brought to the surface. That's because the pressure around them decreases as they are reeled in, which makes their cells expand until they burst. Tough luck for the fish, but that's Boyle's Law in action.

Or you may have noticed that unopened bags of potato chips puff out like balloons when you drive into high, mountainous regions. Since the atmospheric pressure is less at high altitudes, the volume of air in the chip bag expands.

Like many other discoveries of the Enlightenment period, Boyle's Law seems almost self-evident now, but in Robert Boyle's time it was a real eye-opener.

THE WIFFLE BALL LAUNCHER

Science marches forward but not always in the direction we want it to go. Sometimes things get invented that, really, I wish were not. Choosing the "worst invention of the century" is a pretty subjective exercise no matter how you go about it. Still, some ideas are just terrible no matter how you look at them. In this chapter, we'll cover two crummy ideas (at least in my opinion) and then draw what I think is a valid, if a bit tortured parallel between them. First, let's look at the 20th century's worst idea: the baby nuke.

During the dark days of the Cold War, nuclear war meant one thing—mass devastation on a scale so horrific that the mere thought of it normally stops the conversation right there. But in the middle of the 20th century, army war planners hit on the idea of waging small-scale, limited nuclear war. Thus was born the idea for something called "tactical" nuclear weapons.

The first artillery-based nuclear weapon was the M65 cannon. Nicknamed Atomic Annie, this weapons system fired a 15-kiloton

113

shell about seven miles. Soon after Annie's inception, an even smaller nuclear cannon was developed.

Nuclear Cannon

Code-named Davy Crockett, the small weapons system consisted of a recoilless rifle and small, portable nuclear shell. In my opinion, Davy was a terrible rendition of a terrible idea on many levels. The nuclear payload would destroy most everything within a half mile of where it landed and release a cloud of deadly radioactivity in the process. Since the rifle had a maximum range of only about three miles, the gunners were bound to get a good dose of radiation themselves, especially if the wind was blowing their way.

Worse, it's not hard to imagine that if one side used a small nuke, then the other side would respond in similar fashion. After a few rounds of retaliation and counterretaliation, the big ICBMs would likely start flying, after which cockroaches and tardigrades would emerge as the dominant species on the planet. So that's why I think tactical nukes are the worst idea ever.

OK, enough with the serious stuff. The second worst invention of the 20th century is the leaf blower.

Your local home center has lots of gas- and electric-powered outdoor machines that really can make your life better. While I can't say I love my lawn mower or snow blower, they do serve an important purpose. But the leaf blower? Like the Davy Crockett, it is something that should have never been invented. To subjectively paraphrase Shakespeare, a noisy, consumer-grade leaf blower is a device used "by an idiot, full of sound and fury, and that does almost nothing."

My former neighbor Big Pete loved his leaf blower. For him, any time was leaf blower time. Instead of using a rake or broom to clean up his yard, he'd joyfully shoulder up his leaf blower and spend a merry hour or two shattering the neighborhood peace with a 100-decibel roar, doing a job that could have likely been done in half the time with a manual garden tool.

Not only that, these breathtakingly irrational wasters of energy are highly polluting. The typical leaf blower engine puts out more pollutants per minute of use than a three-ton Ford pickup truck. The average leaf blower's tiny and primitive two-stroke gas engine (in which oil is dumped directly into the fuel tank and from there, spewed into the atmosphere) has a carbon footprint 30 times greater than that of a full-sized diesel truck. It's true that electric leaf blowers are less polluting than gas-powered ones, but they are still noisy and, in my opinion, they work poorly when compared to a regular garden rake.

So is there any reason on earth for a leaf blower to exist? Surprisingly, yes. It makes a terrific Wiffle Ball Launcher.

The Wiffle Ball Launcher will accurately (by Wiffle ball standards) pitch ball after ball for games or practice. Made from $25 of plumbing supplies, some scrap wood, and your leaf blower, this project shows off your DIY chops to their best advantage. A key component is an easy-to-find PVC plastic fitting called a 3 × 3 × 2-inch "low heel inlet elbow."

By a miraculous quirk of fate, the low heel inlet elbow couldn't be better for turning your leaf blower into a Wiffle ball pitching machine.

First, it accepts a 3-inch diameter PVC pipe, which is the optimum diameter barrel for a regulation Wiffle ball. Second, the 2-inch inlet matches up exactly with most round leaf blower nozzles. But here's the magical part: because of the fitting's unique geometry and a fluid mechanics principle called Bernoulli's Law, an object like a Wiffle ball inserted in the top of the fitting will be sucked in and shot out the barrel.

MATERIALS

- 3 × 3 × 2-inch PVC low heel inlet
- (2) Small screw eyes
- (1) 3-inch diameter PVC pipe, 10 feet long
- (1) 3-inch diameter PVC pipe, 18 inches long
- (3) Sawhorses or other sturdy supports
- (1) Bungee cord
- (1) 2-inch diameter PVC pipe, 4 inches long

- Leaf blower (Leaf blower power is described by the number of cubic feet per minute [cfm] of air it discharges. More powerful blowers with higher cfm ratings deliver greater distance and accuracy, so choose accordingly.)
- Duct tape
- Wiffle balls

TOOLS

- Electric Drill
- Twist Drill Assortment
- Safety glasses
- Batter's helmet

BUILD THE WIFFLE BALL LAUNCHER

11.1 Low Heel Outlet Detail

1. Drill two ⅛-inch diameter holes and insert the small screw eyes into the thick part of the low heel inlet, as shown in **diagram 11.1**.
2. Refer to **diagram 11.2** during the remaining steps. Attach the 10-foot-long, 3-inch diameter PVC pipe to the 3-inch vertical opening on the low heel inlet. Don't reduce the length of the barrel because a short barrel won't provide as much range or velocity.

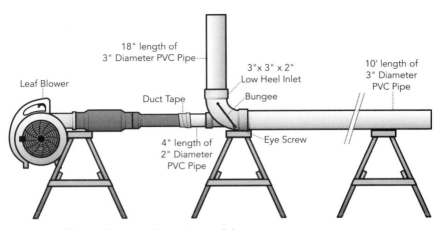

11.2 Wiffle Ball Launcher Assembly

3. Attach the 18-inch long, 3-inch diameter loading tube to the 3-inch horizontal opening on the low heel inlet.
4. Place the barrel, loading tube, and low heel inlet assembly on the sawhorses. Center the assembly and rotate it so the loading tube is vertical. Attach bungee cords from the sawhorses to the screw eyes to hold the loading tube vertical.

11.3 Wiffle Ball Launcher

5. Some air blower nozzles fit well into the 2-inch opening on the low-heel inlet. If yours fits, insert and wrap with duct tape. If not, insert the 4-inch long, 2-inch diameter pipe into the

2-inch hole on the low heel inlet and use reducing fittings as necessary to make it fit, then wrap with duct tape.

6. Align the nozzle of the leaf blower with the 2-inch diameter pipe on the low heel inlet. Use duct tape to securely seal the connection.

BATTER UP!

You're ready to go! Turn on the leaf blower, insert the Wiffle balls in the loading tube, and watch them shoot out. My shooting chronograph shows a muzzle velocity of about 50 mph using my moderately powerful leaf blower. You can elevate the barrel with wood blocks to adjust the trajectory of the ball into the strike zone. Because of the Wiffle ball's holes, each pitch will flutter and curve, making the batting challenge extra fun.

Most leaf blowers don't produce enough power to propel a Wiffle ball with a high enough velocity to be really dangerous. Still, Wiffle ball trajectories are very erratic by design, so be aware you'll frequently get hit by a pitch. Wear a batter's helmet and eye protection.

Victory Through Airpower

Air guns, in one form or another, have been around for thousands of years, in fact, for much longer than gunpowder weapons. If you think about it, blowguns are a type of air gun that has been used for hunting since prehistoric times.

The use of mechanically powered air guns goes a long way back in time as well. The earliest air guns were simply blowguns powered by a bellows attached to the breech. Instead of huffing into the pipe, some clever large-game hunter came up with the notion of putting a squeezable bag on the end. When the bag was squeezed, the compressed air shot a dart or pellet out of the gun. And if the bag was squeezed using the mechanical advantage derived from a system of levers (imagine a fireplace bellows), then the gun could be made to shoot much more powerfully than could be accomplished through lung power alone.

The oldest existing mechanical air gun is thought to be a specimen in the collection of the Livrustkammar Museum in Stockholm. Inside this old gun, which the museum dates to about 1580, a spring mechanism operated an air bellows located in the stock of the gun. When the shooter pulled on the gun's trigger, the spring caused the bellows to force a powerful air gush that shot a specially shaped bolt, or dart, toward the target.

By roughly 1600, air-powered darts were being shot across Europe for sport in a variety of ways. According to the people who study such arcana, spring-powered air guns activated by a moving piston (which was a big improvement upon the earlier bellows-reservoir technology) quickly appeared. An early, exceptionally detailed description of such an air gun is found in the Elemens d'Artillerie by David Rivaut, who was preceptor to Louis XIII of France. He ascribes the invention to a man identified only as "Marin, a burgher of Lisieux," who presented the first air gun to England's Henry IV. By the turn of the 19th century, air guns had developed to the point where they were likely more accurate and more powerful than black powder weapons of similar size.

Circa 1800 air guns of any size and quality were frightfully expensive to make. It took months of time, arcane knowledge, and excellent tools to make a device of this type because the components—valves, locks, cylinders, and reservoirs—had to be very carefully machined. Consequently, an air gun cost far more than a simple black powder rifle and was beyond what most people of the time could spend on a sporting piece.

But for those who could afford them, air guns offered a lot of advantages. By comparison with a smoothbore, muzzle-loading musket, air guns were a hunter's dream. For one thing, they could be fired several times a minute, far more readily than the muskets of those days, which required a load-tamp-fire procedure.

Second, they didn't put out a lot of smoke. This made it easier to aim the next shot if it was needed; line of vision wasn't obscured by powder smoke. Third, a shooter didn't need to be concerned about keeping his powder dry; it worked as well in damp weather as in dry.

Probably the most famous air gun in American history was a rifle carried by Meriwether Lewis during the Lewis and Clark expedition of 1803–1806. The actual gun (although there is some controversy surrounding its pedigree) resides in the Virginia Military Institute's museum of historical weapons. The VMI museum claims that the .31″ caliber, flintlock-style pneumatic rifle in its collection is the one built by expert clockmaker Isaiah Lukens in Philadelphia and hauled to the Columbia River Valley and back by the Core of Discovery.

APPENDIX 1

MATERIALS

Each project description provides a materials and tool list to get you started.

Choosing the right materials for your projects is very important. There are thousands of different materials available, from woods to metal to plastic. And of course, there are many different types of woods, thousands of different metal alloys, and just as many different types of plastics. Each material has its own particular characteristics, its own strengths and weaknesses, that make something suitable for one project but not for another.

You're probably familiar with at least some materials already, but we'll delve a bit deeper into them than perhaps you have gone before. I'll also give some advice on the best places to obtain the hard-to-find items.

WOOD

Wood is probably the most common DIY raw material. It's inexpensive, strong, and easily cut and joined. There are two general types of woods: hardwoods and softwoods. Softwoods, like pine and fir, are the type that fill the big racks at the lumber yards. That's the type we'll use in our projects. The common sizes of wood used in construction projects (for instance 2 × 4 or 1 × 6) are the ones that carpenters use, and that type of wood goes by the name "dimensional lumber."

By the way, 2 × 4s are not actually 2 inches by 4 inches. When the board is first rough sawn from the log at the sawmill, it really is 2 inches by 4 inches. But the log is then dried and planed, so the final, finished size is actually 1½ inches by 3½ inches.

If you need wood pieces that are dimensionally accurate, for example, a 1-inch by 1-inch board that really is a full 1-inch × 1-inch square, then find the wood dowel section of the store. Square wooden dowels are sized exactly as marked.

- The projects in this book make use of softwood dimensional lumber, usually pine or fir.
- Buy the wood for these projects at lumber yards or hardware stores. Lumber yards are usually cheaper and have larger selections than hardware stores. Also, many lumber yards will cut boards to size for free or for nominal cost.
- Some projects call for pieces of wood called dowels. Dowels are round or square rods and are sold in 3- or 4-foot lengths at lumber yards and hardware stores.

METAL

Metals make up the world's most important engineering and construction materials, and they are used in nearly every aspect of our lives. Industrially, metals such as steel, copper, and aluminum are used in cars and construction products, refrigerators, and washing machines. And although metalworking is in some ways more difficult than working with wood or plastic, it's often the material of choice for the DIYer as well, especially when the project calls for strength and formability.

- Lumber yards and hardware stores nearly always have modest selections of steel, copper, and aluminum pieces. It is usually found in a self-contained display arranged by metal type and shape.
- Use tool bits, blades, and abrasives designed for use with metal.
- Cut metals often have sharp edges called burrs. Remove burrs with sandpaper or a file before handling cut metal.

PLASTIC

Plastics are polymers: solids made up of large molecules full of repeating chemical bonds. Lightweight and strong, plastic is used in many projects in the book. Most of the projects call for PVC plastic, which is widely available because it is used for home plumbing projects.

- Common PVC plastic pipe and pipe fittings are sold in hardware and home stores.
- There are different types of PVC. Schedule 40 is a good all-purpose plastic. Don't buy non-pressure-rated pipe or fittings for these projects, as it probably won't be strong enough to withstand the pressures involved. If in doubt, look for the words "schedule 40" or the pipe's pressure rating printed on the piece, or ask the clerk.
- To cut PVC, you will find that most hand and power tools will work adequately. If you cut PVC frequently, invest in a ratcheting PVC pipe cutter, which will make the job faster and cleaner. To join PVC pieces together, see the instructions following.

HOW TO SOLVENT WELD

1. First, clean the surfaces to be welded with PVC primer. Apply the primer with a dauber or brush. The primer cleans and softens the PVC. Allow the primer to dry before applying cement.
2. Brush on a coat of PVC solvent cement. Apply plenty of cement, first to the pipe end and then to the fitting socket. Leave no bare spots.
3. Immediately join the pipe and fitting so the pipe is fully seated in the fitting's socket and make a slight twist to distribute the solvent evenly inside the weld. A tiny but continuous ooze of cement around the fitting-pipe joint indicates that you used enough solvent cement to ensure a leak-free connection. Let the joint dry according to the directions on the PVC cement packaging before using.

PVC FITTINGS

There's a special terminology that describes plastic pipe fittings. First, let's talk about the four ways PVC fittings connect to pipes or other fittings.

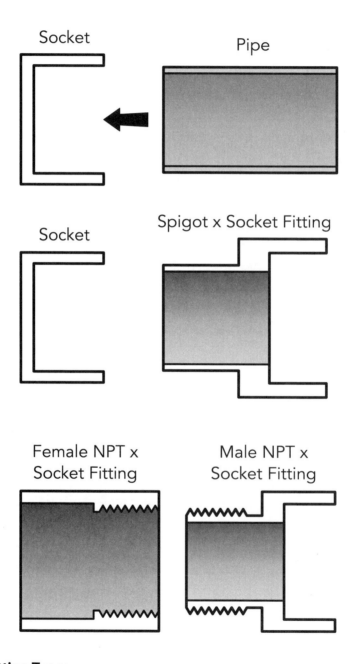

PVC Fitting Types

Slip or Socket Fittings: These are fittings with two smooth female ends, sized to accept standard-sized PVC pipes in their openings. PVC cement or solvent is used to connect pipes into the socket openings. Couplings, tees, and elbows are common socket-style fittings.

Spigot or Street Fitting: A spigot or street fitting has one side that is designed to be inserted and then solvent welded directly into the socket of another fitting. Spigots eliminate the need to cut a separate piece of pipe to connect two fittings together. Bushings and street elbows are typical examples of a fitting with a spigot end.

MPT or MNPT Fittings: MPT and MNPT is shorthand for male (iron) pipe thread. These fittings have external or male screw threads cut into one or both ends. Typically, they will simply screw into a female iron pipe fitting.

FPT or FNPT Fittings: FPT stands for female (iron) pipe thread. These fittings have internal or female pipe threads cut into one or both ends. These fittings are designed to accept a threaded iron pipe.

Most hardware stores carry a variety of PVC fittings. In some cases, you may need to use combinations of fittings to enlarge or reduce diameters to the sizes described here. Consult the hardware store clerk for assistance on selecting fittings. You can also order fittings online from vendors such as:

- FlexPVC: www.flexpvc.com
- PVC Fittings Online: www.pvcfittingsonline.com

SWITCHES, WIRE, AND OTHER ELECTRICAL SUPPLIES

Some electrical supplies and components may be purchased at hardware stores. More specialized items such as project boxes are available at online suppliers that have large selections.

Radio Shack's website provides a reasonable inventory of electrical parts and wire at competitive prices. It's a good place to start. If they don't have what you need, then try Newark Element 14 (www.newark.com). This company has a tremendous inventory, but it takes some knowledge of electronics to figure out what to order.

MISCELLANEOUS TOOLS AND MATERIALS

There are several online industrial supply and distribution companies with very large and comprehensive websites. The companies listed here, while not always the lowest priced, are quite easy to do business with, ship fast, and have huge inventories of parts, tools, and raw materials.

- McMaster-Carr: www.mcmaster.com
- Grainger: www.grainger.com

APPENDIX 2
FIND OUT MORE

If these projects have piqued your interest, here are a few resources for obtaining additional information.

Metal casting is a great hobby, and I've merely scratched the surface in this book. To learn more advanced techniques, I recommend Dave Gingery's slim but information-packed booklet called *The Charcoal Foundry*.

And if you're wondering what daily life was like 350 years later for the sailing men and powder boys in Admiral Nelson's fleet, check out *Men of War: Life in Nelson's Navy* by Patrick O'Brian.

The biggest mushroom clouds of all time were produced during the Cold War in the 1950s. Richard Rhodes's book *Dark Sun: The Making of the Hydrogen Bomb* is both dark and brilliant.

Cannons were used extensively in the American Civil War, and the Battle of Antietam was a frightful example of how powerful they could be. Reading *Landscape Turned Red: The Battle of Antietam* by Stephen W. Sears will make you glad you were not a Civil War soldier.

If you liked making the Mushroom Cloud Cannon, you might want to try making fireworks. A good way to start is with George Weingart's *Dictionary and Manual of Fireworks*. It's an old book, but I've not found a better book to start with.

To learn more about Robert Boyle, Robert Hooke, and the other great scientists whose work provided the foundation for understanding things that shoot, check out my book *The Practical Pyromaniac*.

And if you enjoy making projects of this sort, there are many more to explore in the newly revised edition of my book *Backyard Ballistics*.

INDEX